Selected Formulas (*Continued*)

Term	Symbol	Definitional Formula		
Fisher post-hoc test (unequal N)	t	$t = \dfrac{\bar{X}_1 - \bar{X}_2}{\sqrt{MS_W \dfrac{1}{N_i} + \dfrac{1}{N_j}}}$	158	158
SS value for one-factor between-subject ANOVA error term	SS_W	$SS_W = \Sigma(X - \bar{X}_c)^2$	215	215
SS value for two-factor ANOVA error term for mixed designs				
Between S factor	$SS_{S/G}$	$SS_{S/G} = C_R[\Sigma(\bar{X}_S - \bar{X}_{Gr})^2]$	216	216
Within S factor	SS_{WS}	$SS_{WS} = SS_W - SS_{S/Gr}$	216	216
Correlation coefficient	r	$r = \dfrac{\Sigma Z_x Z_y}{N}$	230	235
Regression line predicting y from x	y'	$y' = b_{yx}X + a_{yx}$	242	242
Slope of regression line	b_{yx}	$b_{yx} = r\dfrac{\sigma_x}{\sigma_y}$	242	242
y intercept	a_{yx}	$a_{yx} = \bar{y} = b_{yx}$	242	242
Standard error of estimate	$\sigma_{y'}$	$\sigma_{y'} = \sqrt{\dfrac{\Sigma(y - y')^2}{N}}$	244	245
Chi-square	χ^2	$\chi^2 = \Sigma\dfrac{(f_o - f_e)^2}{f_e}$	252	259
Phi coefficient	φ	$\varphi = \sqrt{\dfrac{\chi^2}{N}}$	265	265
Cramer's phi coefficient	Cramer's φ	$\varphi = \sqrt{\dfrac{\chi^2}{N(K-1)}}$	265	265
Eta squared	η^2	$\eta^2 = \dfrac{SS_{effect}}{SS_{total}}$	277	277
Omega squared	ω^2	$\omega^2 = \dfrac{SS_{effect} - df_{effect}\, MS_{error}}{SS_T + MS_{error}}$	278	278
Confidence interval for μ	C.I.	$C.I. = \bar{X} \pm t\, S_{\bar{x}}$	281	281

Student:

To help you make the most of your study time and improve your grades, we have developed the following supplement designed to accompany Groninger: *Beginning Statistics Within a Research Context:*

• Study Guide, by Jim Van Haneghan
0–06–042522–9

You can order a copy at your local bookstore or call Harper & Row directly at 1–800–638–3030.

Beginning

Statistics

Within a

Research

Context

BEGINNING

STATISTICS

WITHIN A

RESEARCH

CONTEXT

Lowell D. Groninger

University of Maryland Baltimore County

HARPER & ROW, PUBLISHERS, New York
Grand Rapids, Philadelphia, St. Louis, San Francisco,
London, Singapore, Sydney, Tokyo

1817

Table Credits

Table A   Reprinted from pp. 20–22 of *A Million Random Digits with 100,000 Normal Deviates* by RAND Corporation (New York: The Free Press), 1955. Copyright 1955 and 1983 by the RAND Corporation.

Table B   From *Fundamentals of Behavioral Statistics*, Fifth Edition, by Richard P. Runyon and Audrey Haber. Copyright 1984 by Newbery Award Records, Inc. Reprinted by permission of Random House, Inc.

Table C   Reprinted, with changes in headings, by permission from Pearson and Hartley, *Biometrika Tables for Statisticians*, vol. 1, 1966, pp. 170–175.

Table D   Reprinted, with changes in headings, by permission from Pearson and Hartley, *Biometrika Tables for Statisticians*, vol. 1, 1966, p. 139.

Table E   Taken from R. A. Fisher and F. Yates, *Statistical Tables for Biological, Agricultural and Medical Research* published by Longman Group UK Ltd., London (previously published by Oliver and Boyd Ltd., Edinburgh) and by permission of the authors and publishers.

Sponsoring Editor: Laura Pearson
Art Direction: Teri Delgado
Text Design: Textmark
Cover Coordinator: Mary Archondes
Cover Design: Circa 86, Inc.
Production: Willie Lane

Beginning Statistics Within a Research Context

Library of Congress Cataloging-in-Publication Data

Groninger, Lowell D.
    Beginning statistics within a research context / Lowell D.
  Groninger
      p.   cm.
    ISBN 0–06–042528–8
    1. Statistics   I. Title
  QA276. 12.G76   1990   89–38406
  519.5—dc20   CIP

90 91 92 9 8 7 6 5 4 3 2 1

Contents

Preface

The Problem

Those of us who have taught a course in Elementary Statistics know the problem well. Students often come into the course with great anxiety and a poor background in the essentials needed to understand the material. Then, somewhere around the topics of probability and theoretical sampling distributions, they get overwhelmed and give up on trying to understand conceptually and resort to memorizing procedures. The end product is unsatisfactory for everyone. Students feel that they have spent an inordinate amount of time to gain a very superficial understanding of the material. And their instructors often agree.

The Solution

A.   Philosophy of this Book

Is the problem inherent in the subject matter or can anything meaningful be done? Certainly, the subject matter is difficult for most students, but I believe that most texts greatly compound the problem in many ways. First, I believe that a beginning statistics textbook must have a focus. Comprehensive coverage of all possible topics is guaranteed to overwhelm the vast majority of students. The focus of this book is based on the rationale for requiring our majors to take this course. The rationale is as follows: (a) We would like our students to understand the role of statistics within research. This role includes both their importance and their limitations. The latter particularly needs emphasis. Too often students lose track of the big picture and think that a statistical test is the end product of research or that a statistical test can prove an experimenter's hypothesis. Incredibly, many texts either directly or indirectly teach the latter to students. (b) We would like students to be able to read and understand common statistical tests in journal articles. Since the most common statistical test is the two-factor ANOVA, this means that the statistics course should cover this topic with more emphasis than usual. (c) The course should lay the foundation for individual student research. (d) The course should provide a solid conceptual foundation for those students who will take more advanced statistics courses.

I think that it is important for authors to try to understand *why* students have difficulty understanding and use this analysis in the development of a text. I have studied the problem informally for many years and have used my background in Cognitive Psychology to arrive at a number of reforms that have worked extremely well for me. As I see it, the major reasons that students have difficulty with elementary statistics courses are the following:

1. Information Overload There are too many new things to be learned in too short a time period.

2. Abstract Material and Logic Some of the material is not only abstract, but the student is also required to think in a manner that is totally foreign to him or her. Students have often been taught to think in terms of right and wrong, black and white, not in terms of probability, which is the cornerstone of inferential statistics. Further, the notion that something is likely true because something else is likely false is contrary to their usual thinking style. When these factors are stirred in with theoretical models and math, I think that it is very understandable that students should have problems.

3. Language In some ways this problem is part of the first two problems, but deserves special attention. I estimate that most elementary statistics texts contain between 200 and 400 new terms and symbols. Most of these terms have no synonyms or ''anchors'' with which the term can be associated. Thus, while a definition can be memorized, this definition may have no meaning. Like a parrot, a student may simply learn a string of words. As concepts build upon and give meaning to one another, the language problem compounds. From a student's perspective, it is important to understand that abstract terms defined by other abstract terms are *not* meaningful definitions. There is no substitute for good examples and analogies.

4. Organization The way that concepts and topics are organized within a text interacts with the previously mentioned problems that make things difficult for students. An efficient organization can enormously simplify things.

B. Student Aids

Interesting tidbits and a touch of humor here and there are great, but the real student aid is one that allows the student to understand the material. One key, I believe, is to use a student's knowledge base for positive transfer instead of interference. For example, many statistics texts use the term *observation* to describe a piece of data. To most students, observation is something you do at night when you look at the sky. Thus, negative transfer occurs. I use the term *score* to represent a piece of data because students have had a lifetime of experience with scores representing numbers. A sampling of this and other types of student aids is given below.

1. Examples and Analogies Abstract material is best assimilated into one's knowledge base through tie-ins to established knowledge. The general practice that I have used is to follow abstract terms and concepts with an example or analogy to give it meaning. (See ''A Note on Retaining H_0'' in Chapter 6 for a particularly effective analogy.) Additionally, important definitions are again given in parentheses in later parts of a chapter or in following chapters. Students often need prompts and will ''read over'' terms with which they are only tangentially familiar.

2. Organization of Material A major goal of this text is to teach the basic concepts of inferential statistics. In doing this, I have tried to keep the development of inferential statistics "on line" with as few diversionary topics as possible. Thus, I have pared down some of the descriptive statistics material and graphics (which is not needed as much with modern computer technology) and postponed the topic of correlation until after the inferential statistics story.

I have also maintained a continuity across the most difficult concepts. After probability, I introduce hypothesis testing and decision-making, then immediately follow that with the Sign test which utilizes these concepts. Then, following a presentation of transformed scores and the normal curve, I present the sampling distribution of means followed by a Z-test which utilizes all of the preceding concepts. This is followed by another type of transformation, the *t* value in a *t*-test, which is identical in logic to the Z in a Z-test. All theoretical sampling distributions are introduced by asserting that this type of distribution would arise under thousands of replications of an experiment where H_0 was really true. Thus, students can easily see the logic of using and partitioning a distribution into a 95–5% split.

The ANOVA chapters are developed by emphasizing definitional formulas so that one chapter flows into another, and what many would ordinarily consider the most difficult chapter (i.e., repeated measures and mixed designs) actually involves very little new material. Students mastering Chapters 7 and 8 will have already learned 80% of Chapter 9. There is also an emphasis on interpreting ANOVA. Without direction, students will have no idea of how to relate ANOVA results to tables and graphs or to investigators' predictions.

The role of statistics within research is laid out in the first chapter. The last chapter serves to review, summarize, and broaden the interpretative scope of statistics and research.

3. Simplification Wherever possible, I have tried to simplify without destroying the integrity of the material. For example: (a) I have discussed directionality of tests, but have taken it out of the decision-making process for students. They are advised to use two-tailed tests. This also allows me to omit H_1 from the decision-making process. In this book, retaining H_1 would always be the same as rejecting H_0, so I have omitted the redundant H_1. (b) Distinctions (with one exception) are not made between capital and small letters when used as symbols. One or the other is used, but not both to represent different things. Descriptive subscripts are also used. (c) Some difficult concepts which cause interference are delayed until the difficult concepts involved in inferential statistics have had a chance to consolidate. Thus, there is a separate chapter dealing with power, confidence intervals, and effect size. However, these topics could be integrated into the regular chapters if so desired. (d) Calculators programmed to compute standard deviations can be purchased for about $10 at this writing. With these calculators, ANOVA problems can be done directly from definitional formulas, eliminating the need for complex "raw score" formulas and saving many hours of number crunching. In addition to standard "raw score" formulas, I have included procedures for students to compute ANOVAs with "statistical calculators." (e) In this book, all problems use only the .05 alpha level. (f) Tests where the population parameters are known have been omitted. They are uncommon in applied use and cause confusion for students. (g) Computational procedures are laid out in step-by-step format with examples. Students are "walked through" these examples in the body of the text. (h) I have included chapter outlines, chapter summaries, lists of key words, summary tables and boxes, flow charts where useful, and some notes and tips to the student. There is also a marginal glossary.

4. Active Learning Throughout the text students are asked to consider or "guess at" certain situations. In particular, students have difficulty associating concepts with data. Particularly helpful with this problem are homework problems designed to integrate the data with concepts (e.g., make up 20 numbers that would create a bimodal and symmetric frequency distribution). An extremely useful homework problem that creates a solid conceptual foundation for the relation between $S_{\bar{X}}$ and $\sigma_{\bar{X}}$ is problem 5.15. After doing this problem, students understand this relationship. Some homework problems in each chapter foster a conceptual understanding beyond working with procedures. In some cases, students "discover" principles previously given in the text.

5. Efficiency For efficient learning and review, the writing is concise.

Evaluation

So, do these ideas work? Based on student evaluations, most definitely yes. They have been superb based on photocopied drafts of the text that I have been using for three years. In fact, the book itself has been made immeasurably better from detailed chapter-by-chapter evaluations that students have made. This book owes much to those students. Thanks, guys. I realize that there is more to a text than student ratings. But, on the other hand, if students can't understand a text, nothing else really matters.

Acknowledgments

I would like to thank the following individuals for their useful suggestions concerning the content of this book: Gail A. Bruder, SUNY at Buffalo; Charles S. Weiss, College of the Holy Cross; Roger L. Thomas, Texas Christian University; Geoffrey Keppel, University of California-Berkeley; Philip J. Bersh, Temple University; James Van Haneghan, Vanderbilt University; Keith Stanovich, Oakland University; Susan Dutch, Westfield State College; William L. Kelly, San Diego-Mesa College; Warren Faas, University of Pittsburgh at Bradford; James L. Pate, Georgia State University; Robert Levy, Indiana State University; Heather Resnick, Roosevelt University; Stephen A. Truhon, Winston-Salem State University; Michael D. Biderman, University of Chattanooga; Charlotte Christner, University of

Wisconsin-Whitewater; Carlla S. Smith, Bowling Green State University; Nancy J. Skilling, University of Minnesota; R. S. Melton, University of Cincinnati; Mary Allen, California State-Bakersfield; Steve Falkenberg, Eastern Kentucky University; John B. Best, Eastern Illinois University; Thomas A. Widiger, University of Kentucky; Eric F. Ward, Western Illinois University; Susan A. Warner, University of Arizona; Richard J. Harris, University of New Mexico; Aaron Brownstein, University of North Carolina at Greensboro; and John Harsh, University of Southern Mississippi.

Additionally, I would like to thank my colleagues at University of Maryland Baltimore County, Jonathan Finkelstein and Marilyn Demorest. Jonathan made some useful suggestions on some experimental design issues, and Marilyn caught many of my errors and made several astute technical suggestions.

I would also like to thank everyone associated with the book at Harper & Row. Laura Pearson, sponsoring editor, has been enthusiastic, supportive, and a pleasure to work with. She, her assistants, and the production team have been a class act. And, speaking of class acts, I would like to thank my typists, Madelon Kellough and Terri Harold, for their diligence in dealing with some extremely rough drafts and somehow turning out properly subscripted formulas on machinery not designed for such esoteric enterprises. Thanks for a great job!

Additionally, I would like to thank J. B. Yowell for his instrumental role in instigating the project. Finally, I would like to thank my Indiana relatives for their encouragement, support, and generous offer to help me dispense of any excess royalties that might accrue.

Lowell Groninger

Note to the Student

I once had a kitten who thought that she didn't like milk. She would sniff the milk from a bowl, but not drink it. One day when she was sniffing, I pushed her chin into the bowl, and when she licked the milk from her face, she discovered that she liked milk. This is a true analogy for many of my students, and if you are taking statistics as a requirement, hopefully this analogy will be true for you. Even if statistics turns out not to be your favorite subject, there is a very good chance that you will learn things in this course that will be very valuable to you. For example, you will learn the basis for doing and analyzing research which is the foundation for our knowledge in the social sciences. You will also be developing some logical thinking and critical thinking skills which will be very valuable to you in both your future professional and personal life.

Concerning the material itself, my students who have used this manuscript have found it to be very readable and understandable *provided* that you keep up with the pace of the course. Statistical concepts build upon one another so that, if you miss early parts, your understanding of later parts will be incomplete. You will do yourself a tremendous favor if you develop a schedule that will allow adequate time to master each chapter as you proceed through the text. To help you evaluate your knowledge *prior* to exams and to help you study more effectively, a study guide is available.

I hope that you will enjoy and learn much from the book and your course. Good luck.

Lowell Groninger

Beginning Statistics Within a Research Context

Overview
of Experimentation

THE PURPOSE OF BEHAVIORAL RESEARCH

empirical:

through the senses (e.g., empirical investigation depends on looking, listening, etc.)

In its simplest form, *the purpose of behavioral research is to answer certain types of questions through empirical investigation.* By **empirical investigation,** *we mean procedures that depend on observations to produce ''real-world'' data, as opposed to data that might be generated from mathematical formulas or models.* These data are then used to make decisions. For example, a university may wish to determine the dropout rate for freshmen and, if the rate is high, institute a program to attempt to lower this rate. Although this is a form of empirical investigation, the focus of this book is on a different type of empirical

experiment:

procedure used to establish a causal link between variables

investigation, called *experimentation*. Thus, to understand behavioral research, we will need to discuss what constitutes an experiment, what types of questions are researchable, and what constitutes an answer.

First we elaborate on this statement, and then we explain the important elements in it. First, what is an experiment? You might read in an advice column in a newspaper about a situation involving a teenager experimenting with sex. Does this example concerning the teenager qualify as an experiment in the same way as might a study on the effects of crowding on aggression? At the end of this chapter, you should be able to answer this question.

The second point concerns the type of question that researchers ask. Certain questions fall within the research domain, and others do not. We will establish rules that allow us to decide whether a question is researchable.

The third point concerns determining what constitutes an answer. We usually think of answers as being firm or undebatable, such as the answer to "What is 5 + 5?" or "How do you spell *cat*?" We will see that answers to our research questions are not so simply stated and verified. Instead, answers come about in the context of *probability* and use terms such as "very likely" or "quite confident."

probability:

the likelihood of an event occurring

We will now elaborate each of these points on behavioral research.

WHAT IS AN EXPERIMENT?

variable:

something that can assume different values

independent variable:

variable that is systematically changed, usually by an experimenter

dependent variable:

variable that is influenced by an independent variable

*An **experiment** is a procedure used to establish a causal relationship between variables.* In the simplest form of an experiment, which will be the focus of most of this book, changes in one variable are evaluated to see if they are brought about by changes in another variable. *The variable that is systematically changed is called the **independent variable.** The variable being influenced by the independent variable is called the **dependent variable.*** Typically, changes in the independent variable happen because of a manipulation on the part of an investigator. *A **causal connection** between the independent and dependent variable is established when other variables, which might also influence the dependent variable, are eliminated as reasons for a systematic relationship between the independent and the dependent variable.*

Experiments having one independent variable and one dependent variable can be stated as the effects of A on B. Whatever A represents would be the independent variable, and whatever B represents would be the dependent variable. Manipulations within an independent variable are often established by using different conditions representing different levels of the independent variables. For example, a study on the effects of a drug such as alcohol (the independent variable) on problem solving (the dependent variable) would involve different groups, each group having consumed a different quantity of alcohol. One group may have had no alcohol, a second group may have had 3 ounces of 100 proof, and a third group may have had 6 ounces of 100 proof. After the alcohol had been consumed (over a 20-minute period), each group would be given the same number of identical problems to solve. The dependent variable would be the number of problems solved in a specified time period. Notice that in this experiment the different conditions representing the independent variable involve different amounts of one thing (i.e., alcohol). *The condition having zero amount of the independent variable is called the* **control condition.** *The other conditions are called the* **experimental conditions.** If the number of correct solutions decreased as a function of the quantity of alcohol consumed, we would conclude that alcohol affects problem solving. The validity of this conclusion depends on the elimination of any other (i.e., extraneous) factors that may also have caused the pattern of results obtained in the experiment.

extraneous variable:

a variable that is not a part of the experimental design of an experiment

An example will help to clarify the problem that exists when an extraneous variable influences the dependent variable. Suppose 50 people who consumed no alcohol solve an average of 30 problems within 10 minutes; 50 people who consumed 3 ounces solve an average of 22 problems within 10 minutes; and 50 people who consumed 6 ounces solve an average of 10 problems within 10 minutes. Now, if everything about these three groups is the same except for the quantity of alcohol consumed, then when the experiment is finished we can say with confidence that alcohol impedes the ability to solve problems of the type used in the experiment. However, suppose that half of the people in the 3-ounce group and all the people in the 6-ounce group were members of the school's football team, and the others in the experiment were randomly selected students not involved with the football team. Suppose further that the school's admission requirements were combined verbal and quantitative SAT scores of 1100 for regular students and a bench press of 400 pounds for football players. Thus, although some football players may have SAT scores well above 1100, as a group one would expect them to average well below 1100, since the SAT selection criterion did not apply to them. (This is not meant to pick on football players. Any other group that might be admitted on the basis of a criterion irrelevant to SAT scores would also be expected to average below 1100 on their SAT scores.) Under these conditions, we obviously have an alternative explanation for the results of the experiment. Perhaps it was not differences in alcohol consumption that was creating differences in problem solving. Instead, differences in problem solving may have been due to differences in the extraneous variable (e.g., type of person solving the problem).

When an extraneous variable (i.e., a variable not controlled by the experimenter) covaries (varies along with) with the independent variable, we say that the experiment is **confounded.** An equivalent, but less formal definition of confounding is the following. An experiment is *confounded* if either of two situations occur: (a) an extraneous variable is present in some conditions of the independent variable, but not in other conditions; or (b) the extraneous variable is present to a greater degree in some conditions of the independent variable than in others. When an experiment is confounded, we cannot say whether the results of the experiment are due to the independent variable or to the extraneous variable(s), or perhaps both. It is important to design the experiment in such a way that there are no confounding variables, thereby establishing a causal connection between the independent variable and the dependent variable.

An independent variable can also involve conditions that cannot be quantified beyond noting that the conditions are different. An example of this situation would be a study on the effects of different learning instructions on the number of words recalled from a list of 30 words. Suppose that two sets of instructions for learning a list of 30 words were: (a) make an image representation of each word, and (b) think of a verbal associate to each word. In this situation, we cannot say that one condition has twice (nor any other multiple) the amount of the independent variable as the other condition. It is possible to differentiate these two situations involving the independent variable by using the term *levels* when conditions of the independent variable can be given a meaningful numbered value and by using the term *groups* when conditions of the independent variable cannot be given a meaningful numbered value. However, we will follow the more customary procedure and use the terms interchangeably. Box 1.1 presents a visual display of these two examples. Within this display, each box (more formally called a cell) represents a condition of the independent variable. Scores representing the dependent variable would be placed within each box (i.e., cell). This type of display will be used throughout this book and will greatly facilitate your understanding of more complex experiments.

The type of experiment shown in Box 1.1 is sometimes called a "true" experiment. However, my preference is to use a term that is not so exclusionary. Therefore, we will call this type of experiment a *prototype experiment.* Prototype experiments have the following properties:

(a)   One or more independent variables having two or more conditions per independent variable.

confounded experiment:

an experiment that has results that can be explained by something other than the independent variable manipulation

level:

condition of an independent variable that can be given a meaningful numbered value

group:

condition of an independent variable that cannot be given a meaningful numbered value

prototype experiment:

an experiment that has (a) one or more independent variables, (b) a measurable dependent variable, (c) random assignment of subjects into conditions of the independent variable

random:

by chance factors only

(b) A measurable dependent variable.

(c) Random assignment of subjects into the conditions of the independent variable. ***Random*** *means "by chance factors only."*

BOX 1.1
Two Examples of Prototype Experiments

Example 1. The effects of alcohol on problem solving

Alcohol Level

None	Low	High
x_1	y_1	z_1
x_2	y_2	z_2
.	.	.
.	.	.
.	.	.
x_n	y_n	z_n

The dependent variable is the number of problems solved in a specified time period. The variable x_1 stands for the first subject's score within the "none" condition, x_2 stands for the second subject's score, and so on, down to x_n, which represents the last subject's score within that condition. The y and z scores follow the same pattern for the conditions that they represent.

Example 2. The effects of learning instructions on list recall.

Learning Instructions

Imagery	Associations
x_1	y_1
x_2	y_2
.	.
.	.
.	.
x_n	y_n

The dependent variable is the number of words correctly recalled from a list.

MEASUREMENT SCALES

quantified:

to be measured with
numbered values

Before a prototype experiment can be performed, the dependent variable must be **quantified.** *That is, the variable must be made capable of being measured by numbers that meaningfully reflect the dependent variable.* For example, in the alcohol and problem-solving study, we measured problem solving by the number of problems solved in a 10-minute period. In the learning study, the dependent variable was measured by the number of words recalled from the 30-word list. Numbers can reflect any of four different measurement scales, which in turn can determine how one analyzes and interprets the data from an experiment. The four scales are nominal, ordinal, interval, and ratio.

NOMINAL SCALE

scale (nominal):

form of measurement
that only labels (e.g.,
numbers on uniforms)

*A **nominal scale** names or labels things and is used only to distinguish between things.* With this scale, we can determine that two things are different, but we cannot rank the two things or say how much different one thing is from another. Numbers on the backs of football jerseys is an example of a nominal scale. A player wearing the number 80 is not necessarily twice as good, nor even better, than the player wearing number 40. The only thing that we can say from these numbers is that the players are different. Another example of a nominal scale is numbers on lottery tickets. Their only function is to make each ticket unique. In an experimental context, different sets of instructions representing conditions of an independent variable often constitute a nominal scale. For example, if in teaching spelling to children, we compared a ''repetition-only'' spelling technique with a phonic spelling technique, we cannot say that one technique has more or less of something than the other technique. We can say only that the techniques are different. Therefore, these conditions would represent a nominal scale. Because it can differentiate only between things, the nominal scale is the weakest form of measurement scale.

ORDINAL SCALE

scale (ordinal):

form of measurement
that can rank but not
represent equal intervals

*An **ordinal scale** is capable of distinguishing things and ranking them on some dimension.* However, it does not allow one to talk about the magnitude of differences between ranks. An example of an ordinal scale is the rankings by a sportswriter of the basketball teams in a conference. If all we see are the rankings, then we know that the highest-ranked team was judged the best, but we don't know how big the difference was between the top team and the second team, or even between the top- and bottom-ranked teams. The top team may be head and shoulders above the others (no pun intended) or marginally better than the second team. Similar comparisons can be made for each of the other teams with respect to the others in the group. Note that rankings only order members of a group with respect to each other. They do not say how these rankings can be interpreted on an absolute scale. For example, given five TV shows to rank, a person may do so, but not particularly like his or her top-ranked show. It was simply the best of a bad lot.

INTERVAL SCALE

scale (interval):

form of measurement that has equal intervals but no true zero point

An *interval scale* has all of the properties of nominal and ordinal scales. Additionally, it has *equal distances between each adjacent number in the scale, but no true zero point.* Therefore, it cannot be used to make ratio or percentage statements. An example of an interval scale is the Fahrenheit temperature scale. An increase of 10° is the same amount of increased heat whether the temperature is raised from −5° to +5°, 30° to 40°, or 90° to 100°. Also, since the zero point of a Fahrenheit scale is arbitrary rather than representing zero heat, we cannot say that 80° is twice as warm as 40° or that 10° is 10 times as warm as 1°. There are many scales used in psychological tests that are not truly interval scales, but are usually considered to be interval scales because of their close approximation. Intelligence quotient (IQ) scores and scholastic aptitude (SAT) scores would be two such examples. There is no true zero point on these scales, and the intervals between numbers are designed to be (and are very close to being) equal. Thus, the difference in IQ scores between 90 and 100 would be about the same as between 100 and 110. Also, since IQ scores have no true zero point, an IQ of 140 could not be considered twice that of an IQ of 70.

RATIO SCALE

scale (ratio):

form of measurement that has equal intervals and a true zero point

*The **ratio scale** has all of the properties of the nominal, ordinal, and interval scales,* plus a true zero point. Most of the scales that we are familiar with are ratio scales. All quantities (e.g., gallons, liters), areas (e.g., square feet, acres), time intervals (e.g., seconds, years), distances (e.g., inches, light-years), weights (e.g., pounds, tons), and so forth, are ratio scales. The key to determining whether a scale is a ratio scale or an interval scale is to ask the question, "Does zero amount of whatever it is that I am interested in measuring truly mean nothing?" This is answered in the affirmative for all ratio scales, but not for the interval scale. A memory aid may be helpful in remembering the order of the scales from most powerful to least powerful. The first letter of the four scales, starting with the ratio scale, spells *rion*. The rion is king of the jungle. (I know it's stupid, but it works.)

Sometimes it is not possible to classify a measure as belonging exclusively to one of the four types of scales. For example, rating scales that are bipolar (e.g., ranging from one extreme to the opposite extreme, such as from very good to very bad) are designed to be interval scales. However, since a subjective scale is impossible to verify in terms of equal distances between numbers, an ordinal scale may be the closest to being the correct scale in this example.

Measurement is very important in research. Generally speaking, data that reflect the more powerful scales (i.e., interval, ratio) can give more precise information about what has happened in an experiment. Also, data reflecting interval or ratio scales allow for more powerful methods of analyzing data, as we will discuss in later chapters.

OPERATIONAL DEFINITIONS

operational definition:

states the way in which an independent or dependent variable is measured

Since measurement and numbers are crucial to experimentation, it is of the utmost importance for the investigator to specify how the independent and dependent variables are measured. *Operational definitions state the way in which the independent and the dependent variables are measured.* In our drug study, the independent variable was operationalized in terms of different quantities of alcohol, and the dependent variable was operationalized in terms of the number of problems solved in a 10-minute period. Both measurements involve ratio scales, because zero truly represents nothing and the intervals between numbers are equal. In our second example, learning instructions were operationalized by using two different types of learning instructions (e.g., imagery and associations). Learning instructions would be represented by a nominal scale. The operational definition for the dependent variable would be the number of words recalled within a 5-minute recall period. The scale representing number of words recalled would be a ratio scale.

The scales that represent the dependent variables in our examples reflect performance, not an underlying characteristic of a person. For example, a person may have a score of zero in terms of number of problems solved in a 10-minute period. This score of zero meaningfully reflects his or her *performance* on the task, but may not meaningfully reflect this person's problem-solving *ability*. Typically, the concern of the experiments described in this text will be with the comparison of performance between conditions, not with the measurement of characteristics of people.

QUASI-EXPERIMENTS

quasi experiment:

experiment whose design allows for explanations of the results other than the independent variable

Our major concern in this book will be with what we have termed *prototype* experiments. However, sometimes it is not possible to do a prototype experiment in a situation where one would be desirable. Let us examine some situations where this might be the case. A common aspect of all *quasi-experiments* is that they are inferior to prototype experiments in allowing an inference of causality to be attributed to the independent variable. The major distinction between prototype experiments and quasi-experiments is that there is no random assignment of subjects to conditions with quasi-experiments. The result is the presence of one or more confounding variables in these designs. The purpose of introducing you to quasi-designs is to prevent you from forming the false impression that the only type of design used in psychology and the social sciences is the prototype design. Quasi-designs have niches in many research areas, and often the effects of the confounding variable or variables are minimal, in which case you can have confidence in the results from these designs.

PRE-POST DESIGNS

design (pre-post):

design where a measurement is taken before and after an event has occurred; differences are attributed to the event

With *pre-post designs,* data are collected before an event occurs and also after the event has occurred. If changes occur, the event becomes a likely reason for these changes. Political polls are a good example of this type of situation. A president's popularity rating before a major event involving a president's decision is often compared to his rating after the event, the purpose being to gauge the public's reaction to the handling of the event. A crisis involving a foreign power is an example. Another example is the public's reaction to nuclear arms after viewing a TV movie depicting a nuclear war.

With this type of design, time (or, more accurately, processes that occur in time) is confounded (i.e., also varies) with the event that occurred between the pretest and the posttest. That time may be important is illustrated by the following example. Suppose Brand X company puts on a big advertising campaign, claiming that 90% of the people who take Brand X aspirin get rid of their headaches. The implication being made by Brand X company is that their aspirin is the causal agent. However, they are leaving out an important fact. What percent of the people would have lost their headaches over a given time even if they had taken nothing? This percentage might account for most or perhaps all of the 90% recovery rate.

STATIC GROUPS AND SELF-SELECTED GROUPS

design (static):

design whereby subjects are placed into groups by a common characteristic

With *static groups* and *self-selected groups* designs, subjects are not randomly assigned to conditions from which comparisons are made, but are placed into groups by a common characteristic (i.e., static groups) or by self-selection. Since subjects are not randomly assigned to conditions, explanations other than the independent variable are possible with these designs. As an example of static groups, suppose we hypothesize that a crowded environment creates a higher crime rate than does a spacious environment. To test this hypothesis, we compare the mean crime rate of the five most densely populated cities in the United States with the mean crime rate for five randomly selected cities with low density. Suppose we find a much higher crime rate for the densely populated cities. Does that mean that the hypothesis is correct? *No,* not unless other possible explanations resulting from confounding variables can be eliminated. In the described situation, there are several differences between high- and low-density cities that might account for the difference in crime rates. For example, there may be differences in income, neighborhood stability, religious commitment, number of police officers per capita, and so on. Thus, from this type of design, we cannot say with much certainty that crowding was influencing crime rates.

An example involving *self-selection* is the following. Suppose a teacher has a class size of 40. She wishes to find out if a new handout prepares students better for a test than the old handout does. She therefore asks for volunteers to study from the new handout, and obtains 20 volunteers. The other students are given the old handout. If she finds that students studying from the new handout do better on the test, can she conclude that the handout was the reason? The problem is that the groups are self-selected. The volunteer group may have been brighter, more motivated, or had some other helpful trait, and thus did better because of one of these factors instead of as a result of using the handout. This does not mean that the new handout was not better; it only means that we cannot be sure of this because of other possible reasons. Notice how a prototype experiment would have solved this problem. With random assignment of students to handouts, we can be very confident that the groups were closely matched in terms of ability, motivation level, and any other subject variable. In fact, we can place precise probability levels on our confidence of equivalence between groups with random assignment of subjects to groups. This is an important point, because *probability based on random assignment is the starting point for all the statistical tests of inference that we will discuss in this book.*

BASELINE OR TIME SERIES DESIGNS

design (baseline or time series):

design wherein several measurements are taken before and after an event or manipulation has occurred; an attempt is made to establish a stable baseline before the independent variable (i.e., event) is introduced

We will use the terms *baseline* and *time series* interchangeably. This design is similar to the pre-post design but with two important differences. First, several measurements are taken both before and after an event or manipulation has occurred. Second, if possible, an investigator waits until a stable baseline is reached before introducing the independent variable. This type of design is often used when a single organism is the subject for an experiment. For example, suppose one wished to test the effects of an attractive member of the opposite sex on blood pressure. Assume a laboratory situation where people are coming and going. Once a stable baseline has been established, the attractive opposite-sex person appears briefly, then disappears. If the blood pressure of the subject goes up when the attractive person appears, this is evidence for a causal connection. This evidence can be made stronger if, after a return to baseline, the attractive person is reintroduced with the same results. When an independent variable manipulation is followed reliably by a consistent baseline change, a strong causal connection can be made. This approach generally does not use the inferential statistics models that are the focus of this book. Therefore, we will not pursue this type of design further.

WHAT IS AN ANSWER?

hypothesis:

a statement that can be true or false

The term *answer* is not commonly used among scientists since it implies a final and unequivocal resolution of a problem. Instead of using the terms *question* and *answer*, scientists use *hypothesis* and **confirmation of hypothesis**. *A **hypothesis** is a question placed in statement form that can be true or false* (e.g., I hypothesize that children raised in foster homes will have a higher divorce rate than children not raised in foster homes). To confirm a hypothesis means that the hypothesis is very likely true. *The confirmation is established by showing that the hypothesis is consistent with experimental evidence from a well-controlled design, whereas this experimental evidence is inconsistent with competing hypotheses.* Confirmation of a hypothesis, then, is very closely related to issues involving experimental design and experimentation.

valid measure:

a measure that measures what it is supposed to measure

Whether it is possible to confirm a hypothesis depends on our ability to give a valid measure to two or more variables that we are interested in. In the prototype experimental design, this means giving valid operational definitions to the independent and dependent variables. In a nutshell, **if you cannot measure something, you cannot meaningfully research it.** The measurement scale can range from nominal to ratio, but some validity must be attached to the measurement. The next question is, "What makes a measurement valid?" This issue is very complex and beyond the scope of this book, so we give a working definition: *A measurement is valid when knowledgeable people in the area of what is being measured agree that the measurement is valid.* Some examples will help clarify this.

Example 1. Does talking to plants make them flower sooner? This question is researchable (whether a question is researchable has nothing to do with how interesting or how important it is). The question is researchable because we can meaningfully measure the potential independent and dependent variables. The independent variable would be the presence or absence of speech (the amount of speech would be represented by a ratio scale) during the care of the plants. The dependent variable would be the time (a ratio scale) that it takes a plant (of a given maturity) to bloom.

Example 2. Are there more Democrats in heaven than Republicans? This is not a meaningfully researchable question because it is highly unlikely that agreement exists among a broad spectrum of religious people, or any classification of people, as to by what criteria people go to heaven. Notice that an arbitrary operational definition can be given to going to heaven (e.g., all people attending 17 church services go to heaven). However, if this definition is not accepted by others, then it has no validity and, therefore, conclusions reached by the researcher will not be accepted.

One other point can be illustrated from this example. It is much easier to establish the validity of an operational definition if the variable that we are defining can be directly observed. The most convincing case in our politician and heaven example would be if heaven had regular visiting hours and easy accessibility. One could then visit heaven, verify who was there (assuming a directory of some sort), and come up with some convincing data with respect to the hypothesis.

However, meaningful operational definitions can be given to some abstract concepts. Motivation in the context of school might be measured in terms of hours of study. Hunger might be measured by the quantity of food eaten in a specified period. Imagery abilities might be measured by performance tasks involving mentally rotating drawings and figures. The necessary ingredient in giving an operational definition to an abstract concept is to make sure the measure is consistent with the established scientific meaning of the concept. If it is, other researchers will likely accept the operational definition as being valid.

A second necessary ingredient in confirming a hypothesis is the following. If the hypothesis requires an independent variable manipulation, then it must be possible to make this manipulation for the hypothesis to be confirmed. For example, one might hypothesize that the genes responsible for Huntington's disease could be rendered impotent if they were altered at the appropriate time. This hypothesis is not testable unless the responsible genes can be located and altering techniques become possible.

RESEARCH AND DATA ANALYSIS

Let us assume that we did an experiment involving jury simulation. Our hypothesis is that college students will give longer sentences to the perpetrator of an armed assault if the assailant came from a high-income family as opposed to an impoverished family. We therefore write a description of an armed assault and robbery. The evidence against the defendant is overwhelming, and he pleads guilty. The story is given to 20 randomly selected college students, who are randomly assigned into 2 groups; 10 are told, as part of the defendant's background, that the defendant came from an impoverished home, whereas the other 10 are told that the defendant came from a wealthy family. After reading the case story, the subjects are asked to sentence the defendant to some number of years in prison based on the information given. The independent variable would be the type of economic background of the defendant, whereas the dependent variable would be the number of years that the defendant was sentenced to prison.

Table 1.1 shows three hypothetical data sets. Before reading further, look at each set and decide whether or not that data support the hypothesis that longer sentences will be given to defendants of higher socioeconomic backgrounds. Then state the reason for your

TABLE 1.1
Three Sets of Hypothetical Data

Group A		Group B		Group C	
Low Income	High Income	Low Income	High Income	Low Income	High Income
10	21	15	16	12	14
12	19	10	9	15	11
7	15	18	21	9	19
1	18	20	17	7	20
11	23	12	11	19	23
9	30	25	26	6	9
10	16	8	17	13	17
8	19	14	6	10	15
2	26	17	14	17	16
6	20	6	8	14	21

decision in each case. The first set of data supports the hypothesis, but the second set of data does not support the hypothesis. The third set of data is too close to call. If you got the correct answer for the first two data sets, you already have a good "feel" for data analysis.

Look again at data set A. The numbers in the high-income background group are all higher than the numbers in the low-income background group. Higher numbers mean longer sentences, which is what the hypothesis predicts. Because of the large differences between the numbers of the two groups in set A, there is no problem in deciding that the outcomes within the conditions of the independent variable were markedly different. Another way of saying this is that the independent variable manipulation was *effective*.

Now, look again at data set B. The numbers in the two groups are not exactly the same, but they look, as a group, very similar to each other. Each group has about the same range (i.e., the highest score minus the lowest score), and the mean (i.e., the sum of the scores in a group divided by the number of scores in that group) of the two groups looks to be about the same. To try to argue that differences exist between the two groups based on data set B is obviously going to be unconvincing. If data from experiments always occurred as in sets A and B, data analysis would be very simple, and none of the statistical tests that we later describe would be necessary.

Unfortunately, data often occur as in data set C. Here, there is a problem. The numbers in the high group are generally, but not always, higher than the numbers in the low group. Would we still be justified in saying that a difference exists between the two groups and that the independent variable manipulation was effective? Suppose you say yes, and I say no. How can you convince me, or how do we resolve the issue? The resolution of this issue involves *inferential statistics*, which is the focus of most of the rest of this book.

EXPERIMENTAL PROCEDURE SUMMARY

The components of an experiment are outlined in Table 1.2. An *empirical question is one that directly involves the conditions of the independent variable.* Thus, for example, socioeconomic status is or is not a factor in decisions by juries. *Theoretical questions are one step further removed from the data of an experiment and usually involve an implication of why.* For example, one might start with assumptions that (a) it is worse to commit a crime if one has had an opportunity to make a success of oneself, and (b) a person of low socioeconomic status has not had as much opportunity as a person of high socioeconomic status. Together, these assumptions lead to the prediction that high socioeconomic criminals would receive stiffer penalties. The data from an experiment can directly confirm an empirical hypothesis. However, they can only

TABLE 1.2
An Outline of an Experiment

		Illustrative Example
Step 1	Theoretical question	Does the perceived opportunity for success create a predisposition in a juror with respect to a defendant?
Step 2	Empirical question	Will juries give longer sentences to defendants of high socioeconomic status?
Step 3	Measurement	
	a. How will I.V.'s be chosen	Use defendants of high and low socioeconomic status
	b. Condition of I.V. represents what scale?	Ordinal if determined by someone's ranking
	c. How will D.V. be measured?	Number of years in prison sentenced to defendant
	d. What scale (nominal, ordinal, interval, ratio) will D.V. use?	Ratio
Step 4	Experimental situation	Mock jurors read a description of a crime committed by a person of high or low socioeconomic status. The jurors then give hypothetical sentences.
Step 5	Analysis Type A—Prototype experiment. def.: random assignment of S's to conditions of I.V.'s. Type A uses statistics model. Type B—Quasi-experiment. def.: nonrandom assignment of S's to conditions of I.V.'s. Type B may or may not use statistics model.	
Step 6	Decision a. There is an effect of I.V.'s on D.V.; or b. An effect of I.V.'s on D.V. is not established by this experiment.	
Step 7	a. Inference to step 2 b. (optional) Inference to step 1	

indirectly confirm a theoretical hypothesis, because the reason that the results occurred as they did may have nothing to do with a given set of assumptions. For this reason, we say that a theoretical hypothesis is one step removed from inferences about the results of the data analysis.

Notice that the data analysis portion (step 5) is but one part, although a very necessary part, of the larger picture involving an experiment.

SUMMARY

An *experiment* is a procedure used to establish a causal link between variables. Experiments consist of at least one independent variable (usually manipulated by the experimenter) and a dependent variable. If changes in the independent variable bring about changes in the dependent variable, and if there are no extraneous variables that can account for those changes in the dependent variable, then we can say that the independent variable caused the changes in the dependent variable. When changes in a dependent variable can be explained by an extraneous variable or variables, we say that the experiment is *confounded*.

A *prototype* experiment has the following properties: (a) one or more independent variables having two or more conditions per independent variable; (b) a measurable dependent variable; (c) random assignment of subjects into conditions of the experiment.

The four types of measurement scales are (a) *nominal*, which only distinguishes between things, (b) *ordinal*, which distinguishes and ranks things, (c) *interval*, which distinguishes and ranks things, has equal distances between numbers, but has no true zero point, (d) *ratio*, which has all of the properties of the other scales plus a true zero point. Operational definitions state the way in which the independent and dependent variables are measured. If the independent and dependent variables cannot be meaningfully measured, an experiment using these variables will have no clear interpretation.

Quasi-experiments are similar to prototype experiments, except for an inherent extraneous variable, which may make the interpretation of the experiment less clear. There are three general types of quasi-experiments:

1. Pre-post designs: with these designs, data are collected before and after an event occurs. If differences occur in the predata and post-data, the event becomes a likely reason for these differences.
2. Static groups and self-selected groups: with these types of designs, subjects are not randomly assigned to conditions from which comparisons are made; rather, subjects are placed into

groups by a common characteristic (i.e., static groups) or by self-selection.

3. Baseline or time series designs: this design is often used with a single organism and is similar to the pre-post design with two exceptions: (1) several measurements are taken both before and after the event or manipulation has occurred; (2) if possible, an investigator waits until a stable baseline is established before introducing the independent variable.

Scientists ''answer'' questions by confirming hypotheses. A hypothesis is confirmed by showing that it is consistent with experimental evidence from a well-controlled design while the experimental evidence is inconsistent with competing hypotheses. In prototype designs, a hypothesis cannot be confirmed unless (a) valid operational definitions can be given to the independent and dependent variables, and (b) it is possible to make a required independent variable manipulation.

The role of data analysis in an experiment is to allow the investigator to make decisions about the outcome of the experiment. An *empirical* outcome is directly related to the data in an experiment. A *theoretical* conclusion is one step (i.e., inference) removed from the empirical outcome of the experiment.

KEY TERMS

independent variable

dependent variable

control condition

experimental condition

confounded experiment

levels of an independent
  variable

groups of an independent
  variable

prototype experiment

random assignment

nominal scale

ordinal scale

interval scale

ratio scale

operational definition

quasi-experiment

pre-post design

static groups design

self-selected design

baseline design

hypothesis

confirmation of hypothesis

valid measure

empirical outcome

PROBLEMS

1. Where appropriate, state (a) the independent variable (I.V.), (b) the dependent variable (D.V.), (c) the control condition, (d) the experi-

mental conditions, (e) how the I.V. and D.V. were operationalized, (f) the I.V. and D.V. scale, (g) the confounding variables, if any, (h) the type of experiment (e.g., prototype, baseline, etc.), (i) the hypothesis of the experiment, if one was either stated or implied. If no hypothesis was stated, make up your own.

EXPERIMENT 1

An investigator wishes to find out if listening to rock music impedes the driving performance of teenagers. She gets access to a driving simulator that records the number of errors a driver makes during a 10-minute "run." Three conditions are formed: (a) a no-music condition, (b) a classical music condition, and (c) a rock music condition. The classical music was recordings by Bach, whereas the rock music came from tapes from a radio station known as 44 Rock. The music is played at a 60-decibel level, 5 feet from the subject. The subjects are 30 paid teenage volunteers who are randomly assigned to one of the three conditions. The investigator predicts a deficit in performance for the rock music condition.

EXPERIMENT 2

An investigator wishes to test the effects of erotic pictures on blood pressure. He obtains 40 volunteer college students and randomly assigns them to a neutral picture group or an erotic picture group. The subjects are tested individually, and each is given a booklet containing pictures. The booklet for the neutral group contains men and women modeling fall fashions. The erotic booklet contains pictures of men and women judged by a Boston court to be obscene. Both groups are told that there will be at least one sexually explicit picture in their booklet. The subjects are instructed to look at each picture until they are told to turn the page after 15 seconds. After 3 minutes of looking, each subject has his or her systolic blood pressure taken and recorded.

EXPERIMENT 3

An investigator wishes to test the effects of pain on attentional factors. She sets up a high pain condition, whereby a person has to keep one hand in a bucket of ice water for 2 minutes. Then with the hand still in the ice water, the volunteer spends 30 seconds counting the number of g's shown in a passage displayed on a screen. In the no-pain condition, the procedure is repeated except that the person's hand is in water of about 90° (a neutral temperature). The conditions are explained to paid subjects, and a $5.00 bonus is given to anyone who volunteers for the high-pain condition. The groups are then compared in terms of how many g's they missed from the passage.

EXPERIMENT 4

A therapist believes that a "time-out" procedure will stop 4-year-old David from throwing tantrums. He has the parents record the number of daily tantrums for one week with no interventions. Then the parents are told that every time that David throws a tantrum, he is to be placed alone in his room for 20 minutes. The therapist compares the number of tantrums thrown during the pretreatment week with the number thrown during the third week of treatment to determine whether the treatment was effective.

EXPERIMENT 5

An investigator predicts that men will be less affected by bloodstream alcohol levels than women will. She randomly selects men and women as they enter a singles bar and offers to pick up their bar tab for the evening if they will consume three drinks (each drink containing 1 ounce of 100% alcohol) in 20 minutes. Everyone that she asks volunteers to be a subject. Ten minutes after consuming the alcohol, each subject is asked to walk along a beam that is 3 inches wide, 3 inches high, and 20 feet long. The investigator counts the number of times each subject falls off the beam.

EXPERIMENT 6

To evaluate the success of therapy on a psychotic patient, a therapist records the number of nonsense statements made by the patient during each of 30 sessions. Therapy is given only in the last 25 sessions. The therapy is evaluated by comparing the number of nonsense statements made during the first five sessions versus the last five sessions.

2. Determine whether the following hypotheses are testable and state your reasons.

   a People born with no genetic defects are faster learners than people born with genetic defects.

   b Phone numbers can be retained longer if they are visualized.

   c Ingesting LSD will cause hallucinations.

   d It is better to be lucky than talented.

   e People who take at least one statistics course are better lovers than people who have taken no statistics courses.

3. Make up one hypothesis that you think is testable and one that is not. Give your reasons.

CHAPTER 2

Statistics and Frequency Distributions

WHAT ARE STATISTICS?

The field of statistics involves ways of organizing, summarizing, or manipulating sets of data such that we may adequately describe a set of data and/or make an inference (i.e., draw a conclusion) from it. Statistics are our friends. They make life easier for us. They create order out of chaos, and they help us make intelligent guesses about things of which we are not sure. Statistics are useful to the scientist and to the layman. Statistics aid researchers, in many types of studies, in determining whether an independent variable manipulation has been effective. Statistics also allow you to estimate the probability of your getting a certain letter grade in a course. The government uses statistics to determine how much tax you pay and where the money goes. Perhaps without being aware of it, couples use statistics in determining whether they can afford a house and in planning a family (e.g., what percent of the time is a particular birth control method effective?). Knowing statistics can help you make better arguments and not knowing statistics may allow you to be easily deceived. To summarize, statistical knowledge is not just an abstract entity that is likely to sit on the shelf for the rest of your life; rather, statistical knowledge can be very useful to you in many facets of everyday life.

The field of statistics can be broken into two broad categories: descriptive (the type you likely have some familiarity with) and inferential.

DESCRIPTIVE STATISTICS

descriptive statistics:

area of statistics whose purpose is to describe a set of numbers in some summarized form

Descriptive statistics describe a set of numbers in some summarized form. For example, suppose an instructor wants to report the outcome of an exam that had been given to 100 students. If the instructor chooses not to use descriptive statistics, she may simply read or present a handout containing all the scores. If you were in this class, you would likely be disappointed at not receiving some summary statistics. Suppose your score was 81. What does that score mean? If you compare your score to every other score, you have a time-consuming task, and if you don't do your own summary by tallying scores that are higher and lower you will likely forget much of the information about the other scores. Notice how much easier and clearer it is for you if the instructor gives you a useful descriptive statistic such as the *mean* (i.e., the sum of all of the scores in a group divided by the number of scores in the group). If the *mean* were 70 and your score 81, you would have a fair idea of how you did. You can get an even better idea if the instructor tells you the top score or scores. Thus, most of the useful information contained in the 100 scores can be described with two pieces of information, the mean and a measure of score spread. The various types of descriptive statistics are the focus of this chapter and the next.

mean:

measure of central tendency represented by $\Sigma X/N$

INFERENTIAL STATISTICS

inferential statistics:

area of statistics whose purpose is to infer characteristics of a population based on samples from that population

*Inferential statistics infer characteristics of populations based on samples from those populations. A **population** is a group of things, individuals, or measurements that share at least one common characteristic.* In the behavioral sciences, the measurements (i.e., scores) usually come from characteristics of organisms and, in an experiment, constitute the dependent variable. Thus, in Chapter 1 we had populations involving problem-solving scores, test scores, and years-sentenced-to-prison scores, among others. A population of scores may already exist (e.g., the height of all members of the U.S. Army) or potentially exist, although not currently measured (e.g., number of words correctly spelled from a 50-word list by 21-year-old U.S. citizens). A population can be of any size from zero to infinity (i.e., larger than any given number). The zero case would not likely be of interest.

Populations and Samples

population:

a group of things, individuals, or measurements that share at least one common characteristic

*A **sample** is a subset of a population.* Why are samples important? Why not simply use all of the scores that one might be interested in, instead of using a subset? To illustrate, suppose the instructor in our previous example gave you the mean score for the exam based on the first five scores for the exam that were in her grade book. You might correctly reason that the mean of all 100 scores might be considerably different than the mean of the five scores taken from the group of 100. You would want the correct mean, not an estimate. In fact, in situations involving grades, samples rarely play a meaningful role. Unfortunately, quite a different situation exists involving scores obtained from experiments. Here samples are critical for two reasons. First, the total population of scores may not be available. For example, it is usually impossible to have everyone in the population that you are interested in participate in your experiment. Second, in populations small enough so that an experimenter might get a score from every member, it might be very difficult to get every member of that population to participate in the study. The result is that data from experiments are usually samples and not populations.

Representative Samples

sample:

a subset of a population

How closely the characteristics of a sample (such as the mean) match the characteristics of a population depends on two things: (a) the representativeness of the sample and (b) the size of the sample. There are two common ways of making a sample representative: (1) *stratified sampling: selecting members of the sample by defining important attributes of the population and then selecting samples that have these attributes in the same proportion as in the population.* For example, suppose that the population was scores on an attitude test taken by college students. If year in college were considered an important characteristic of this population and if there were about equal numbers of

samples (random):

a sample that has been selected by a method such that each member of a population has an equal chance of being selected for any designation or grouping

sample (stratified):

sample selected such that important attributes of the population are in the same proportion in the sample

students in all four year levels, then one would stratify a sample by putting equal numbers of freshmen, sophomores, juniors, and seniors in the sample. (2) *random sampling: selecting by a method such that each member of a population has an equal chance of being chosen for any designation or grouping.* A typical way of selecting random samples is to use a random number table where all sequences of numbers are random arrangements. These random sequences can be obtained by going in any direction from any starting point. The starting point can be obtained by closing one's eyes and placing the index finger down on a spot within the random number table. From that spot, go down, say, the desired number of digits. For example, if we want to choose 15 people at random from a group of 90, we could assign each person a number from 1 to 90. Then we could use a random number table by putting our finger on a column and reading down that column for 15 sequences, each sequence containing two digits. These two digits would range from 00 to 99 and thus contain all members of our population. The people represented by our 15 numbers would be a random selection. (If any number in our selection occurred more than once, we would continue selecting until 15 unique numbers were drawn.)

Random number tables are also used to make **random assignments** of people or things into groups. For example, suppose we have 15 people that we wish to randomly assign to three groups. Using a random number table, we could do this by first assigning the numbers 1, 2, 3 to condition 1, the numbers 4, 5, 6 to condition 2, and the numbers 7, 8, 9 to condition 3. We would then pick a row of random numbers from a random number table. We would then line up our subjects in a row and one by one assign them to a condition depending on the arrangement of the numbers 1 to 9 in the row of random numbers. For example, if the first four random numbers were 9, 1, 5, 3, the first four subjects would be placed in condition 3, 1, 2, 1, respectively. One would stop placing subjects in conditions when that condition reached its maximum. In this example, that number would be five subjects. This example is but one of many ways that random assignments may be made.

The probability of a large randomly selected sample being representative of a population is quite high. The general principle is as follows: *the larger the random sample, the more representative it is of the population from which that sample was drawn.* The relationship between sample sizes and characteristics of populations is an important one and will be discussed further in Chapter 5.

Terminology

parameter:

a characteristic of a population (e.g., μ)

Characteristics of populations are called **parameters** *and are identified by Greek letters. Characteristics of samples are called* **statistics** *and are identified by italic letters.* Thus, if a characteristic such as the mean of a set of scores refers to a population, it is designated by μ, whereas if

statistic:

a property or characteristic of a sample (e.g., the mean)

it refers to a sample it is designated by $\overline{X}$. A helpful way to remember this is that the first letters are the same: population and parameter, sample and statistic.

Generalization

In experiments, one often generalizes (i.e., extends) the results taken from sample means to the larger population. Based on our discussion of samples and populations, this generalization is valid only for the specific population from which each member had an equal chance of being selected in the random sample. For example, suppose that I do an experiment testing the effects of two different learning strategies on the learning of a 30-word list. I randomly select from volunteers in an introductory psychology course and randomly assign them to the two conditions. To what population can I generalize the outcome of the experiment? The specific population from which I randomly sampled was volunteers from the introductory psychology class at a specific university. Therefore, this is the appropriate population to which this specific sample may be generalized. If I generalize the results of my study beyond the specific population from which I sampled, I do so based on the argument that the population that I used in my study is itself a representative sample of a larger population. In the given example, I would have to assume that the group that I sampled from was representative of college students in general or perhaps adults in general. These are assumptions about generalization that I would have to make that are beyond the definitive generalization that I could make about a specific population based on a random sample from that population.

FREQUENCY DISTRIBUTIONS

Frequency distributions are a means of displaying a set of scores in a form that gives a much clearer picture of how a large group of scores is arranged than if the scores were simply listed. Table 2.1 shows a list of 100 exam scores. Notice how tedious it is to find the high score or the low score or to estimate an average score. With frequency distributions, salient characteristics of a group of scores are much easier to find.

RAW SCORE FREQUENCY DISTRIBUTIONS

frequency distribution (raw score, also called regular):

a summarized count of how many times each score within a set of scores has occurred

*A **raw score** (sometimes called **regular**) frequency distribution is a summarized count of how many times each score (within a set of scores) has occurred.* To construct this type of distribution, we first construct a scale ranging from the lowest score in the distribution to the highest. We then go through the set of scores one at a time and place a tally mark on the scale represented by that number. For 100 scores, we

TABLE 2.1
One Hundred Hypothetical Exam Scores

73	69	77	61	83	65	70	88	89	62
85	63	87	73	77	75	84	81	69	73
66	72	69	71	83	65	67	83	71	75
71	81	67	64	84	69	71	77	75	73
79	69	73	81	64	82	76	69	69	80
66	70	71	84	73	77	75	85	70	83
88	81	87	71	63	70	71	79	63	77
72	75	79	83	66	74	68	77	84	81
81	73	82	75	83	72	66	63	87	74
77	78	74	62	73	78	84	73	79	67

would have 100 tally marks distributed across the scale. A raw score frequency distribution is a display of this count, as shown in Table 2.2 for the data in Table 2.1.

CUMULATIVE FREQUENCY DISTRIBUTIONS

frequency distribution (cumulative):

a frequency distribution that cumulates scores going from low to high on a scale

As the name implies, a ***cumulative frequency distribution*** *cumulates or adds frequencies as one goes up the scale from low score to high score.* To construct a cumulative frequency distribution, we first construct a regular frequency distribution. Then, for each number in the scale, starting from the lowest number, we add the frequency count for that number to the frequency count for all numbers below it. The cumulative frequency for the highest number on the distribution will always be the same as the total number of scores in the distribution. (See Table 2.2.) Notice that as we cumulate frequencies up the scale from low score to high score, the cumulative frequency at each score along the scale is the *frequency* of that score plus the *cumulative frequency* of the next smaller score.

PERCENTILES

percentile:

a number at or below which a certain percentage of ranked scores fall

We can also describe data within a raw score frequency distribution by using percentiles. *A **percentile** is a number at or below which a certain percentage of ranked scores fall.* For example, if we have 40 scores, the 10th percentile would be the 4th lowest score, the 25th percentile would be the 10th lowest score, the 50th percentile would be the 20th lowest score, the 75th percentile would be the 30th lowest score (or the 10th highest score), and so on. The value of a percentile for any given data set can be found by writing the percentile as a percentage and multiplying by the total number of scores in the set. The score resulting from that procedure will be the percentile in question. For example,

TABLE 2.2
A Raw Score Frequency Distribution and a Cumulative Frequency Distribution
for the Data in Table 2.1

Score (X)	Frequency (f)	Cumulative Frequency (cf)
89	1	100
88	2	99
87	3	97
86	0	94
85	2	94
84	5	92
83	6	87
82	2	81
81	6	79
80	1	73
79	4	72
78	2	68
77	7	66
76	1	59
75	6	58
74	3	52
73	9	49
72	3	40
71	7	37
70	4	30
69	7	26
68	1	19
67	3	18
66	4	15
65	2	11
64	2	9
63	4	7
62	2	3
61	1	1

the 60th percentile of a set of 50 scores would be the $.60 \times 50 = $ 30th lowest score. Since we have 100 scores in Table 2.2, the cumulative frequency column is also a percentile column. This is an unusual situation that occurs only in data sets having 100 scores. Note that in Table 2.2, the 50th percentile is 74, because it occupies position 50 from the bottom of the rankings. Because 74 also occupies positions 51 and 52, it also represents the 51st and 52nd percentiles.

interpolation:

a procedure for determining fractional values between intervals

Suppose one wished to find the 10th percentile in a distribution having 33 scores. The 10th percentile would be represented by the score ranked 3.3 from the bottom. Where fractions of ranks are involved, it is common to round to the nearest rank. Thus, in this example, the 10th percentile would be represented by the third score from the bottom of the rankings.

If more precision is desired, we can use *interpolation.* With this procedure, the fraction involved in the ranking is accounted for. To use interpolation, we (a) find the number represented by the rank when rounded up, (b) find the number represented by the rank when rounded down, (c) take the fraction involved in the ranking (e.g., from our example, the .3 from 3.3) and multiply that by the differences between the score from (a) minus the score from (b). In our example, suppose that the third lowest score was 20 and the fourth lowest score was 26. We would have $.3(26 - 20) = .3 \times 6 = 1.8$. (d) We then add the value in (c) to the value in (b), which is the score represented by the rounded-down rank. Thus, the precise number represented by the 10th percentile would be $20 + 1.8 = 21.8$. These rules of interpolation can be placed in the form of an equation as follows:

$$\frac{X}{\text{Interval width}} = \text{fraction} \qquad\qquad 2.1$$

The equation is solved for X, where X is the number in question to be added to the number representing the rounded-down rank. The interval width is the number representing the difference between the scores of ranks when rounded up and rounded down. Fraction refers to the fraction involved in the ranking in question. Thus, in our example,

$$\frac{X}{6} = .3$$

$$X = 1.8$$

The number representing the 10th percentile is therefore $20 + 1.8 = 21.8$.

Table 2.2 has taken the chaos that exists in Table 2.1 and has made order out of it. From the score scale on the left side of the frequency distribution, the top and bottom scores are now apparent. How the scores are distributed (i.e., spread out) along the scale is also apparent. If one looks at the cumulative frequency column, ranking can easily be done. For example, if your score were 80, there would be 27 scores above yours. You would rank 28th out of 100. If your score were 84, you would rank 9th (tied with four others). Rankings are determined by the number of scores above the designated score plus 1.

GROUPED FREQUENCY DISTRIBUTION

There remains one problem with Table 2.2. The scale covers 28 numbers from top score to bottom score. This makes it somewhat difficult to determine just where within this range scores tend to cluster. Also, sometimes the range is so large that presenting a table covering the whole range becomes unwieldy. For example, suppose one constructed a raw score frequency distribution for six exams containing a total of 600 points with scores ranging from 45 to 575. We would have a scale with 530 numbers in it. Or imagine doing a raw score frequency distribution for incomes of parents in a large university where the income range may be from $3000 per year to $3,000,000 per year. Clearly, things can get out of hand. The solution to these problems is to construct a grouped frequency distribution instead of a raw score frequency distribution. *A **grouped frequency distribution** combines scores along the scale ranging from high score to low score into predetermined class intervals* (for example, 61–64, 65–68, 69–72, etc.). The guidelines for constructing such a distribution are:

frequency distribution (grouped):

a frequency distribution using predetermined class intervals along the *x* axis

1. The number of intervals is arbitrary, but the 8–15 range makes for easy interpretation while not losing too much information in the consolidation process.
2. It is desirable, but not always possible, to choose class intervals that are multiples of the total number of whole integers between the high and low scores with the high and low scores included in this range. Computationally this range of numbers is determined as high score minus low score plus 1.
3. Within the guidelines of rule 1, choose class intervals of 5, 10, or other multiples of 5 or 10. When using guideline 3, start the interval at the lowest multiple of the class interval that you are using (i.e., a number ending in 5 or 10). Do not start at the lowest score unless it is a multiple of the class interval.

An example will clarify these procedures. It is not as complicated as it sounds. We will use the scores in Table 2.1 to illustrate.

First, we find the total amount of whole numbers between and including the high and low scores. This is $89 - 61 + 1 = 29$. This does not divide evenly by any number that would yield 8–15 class intervals, so we will use a class interval size of 3, giving us 10 intervals (consistent with guideline 1), as shown in Table 2.3. Notice that we started the class intervals with the lowest score, 61.

To illustrate guideline 3, we construct a different grouped frequency distribution, again from the same data (Table 2.1), using class intervals of size 5 (not consistent with guideline 1). We now have six class intervals, but since we are using class intervals of width 5 we will present the scale starting with a number divisible by 5 (60 instead of 61). This grouping is also shown in Table 2.3. As with any summary, Table 2.3 gives you information that is quicker and easier to see than that in Table 2.2, but, as with all summaries, some detail is lost in the process.

We construct a grouped frequency distribution in exactly the same way as a raw score frequency distribution, except we place the tallies within the class intervals instead of beside a whole number. We can also construct a raw score frequency distribution first and then condense it into a grouped frequency distribution.

TABLE 2.3
Grouped Frequency Distributions for the Data in Table 2.1

Interval Size 3

Score Range	f
88–90	3
85–87	5
82–84	13
79–81	11
76–78	10
73–75	18
70–72	14
67–69	11
64–66	8
61–63	7

Interval Size 5

Score Range	f
85–89	8
80–84	20
75–79	20
70–74	26
65–69	17
60–64	9

GRAPHIC REPRESENTATIONS

A *graph* is a visual aid that uses two axes to display material and allows one to more quickly and easily determine certain characteristics of data. The following are two common graphical techniques for putting scores from a frequency distribution into graphical form.

Histograms

histogram:

displays class intervals from a frequency distribution along the *x* axis of a graph while the height of bars represents frequency counts

A *histogram* (a type of bar graph) *displays the class intervals from a frequency distribution along the abscissa,* or *x* axis (horizontal) of a graph, and frequencies from the frequency distribution are represented on the ordinate, or *y* axis (vertical) part of the graph. Each bar in the graph represents the width of a class interval. The midpoint of a class interval is also the midpoint of a bar. The data in the grouped frequency distribution shown in Table 2.3 (interval size 3) are reproduced again as a histogram in Figure 2.1. Although the data are exactly the same in both cases, notice how the histogram makes certain characteristics of the data stand out. For example, the most frequent score within class intervals is now apparent at the tallest bar. The variability of frequencies across the class intervals (the up and down oscillation of the bars) is also apparent. The shape of the distribution also becomes obvious. (We will shortly elaborate on shapes of distributions and their meanings.) The disadvantage of histograms, as opposed to frequency distribution tables, is a potential loss of precision in determining the exact frequency count for a specific class interval. This frequency count is read from (but really only estimated from) the *y* axis of the graph instead of being specifically given in a table.

Figure 2.1

A histogram of the data in Table 2.3: class interval size of 3.

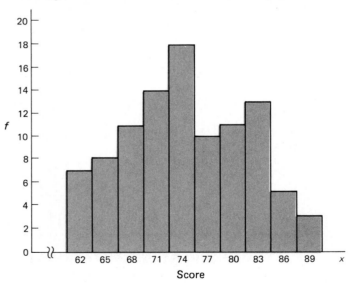

Frequency Polygons

Frequency polygons are line graphs that are particularly appropriate when the data scale represented on the *x* axis is continuous rather than discrete. A ***continuous*** *variable can theoretically assume any fractional value between specific numbers.* Time and distance are examples of con-

frequency polygon:

a line graph representing frequency counts

tinuous variables. Histograms are most appropriate for discrete data. *A **discrete** variable can only assume specific whole number values.* The number of children in a family is an example. Technically, the data in our exam example are discrete. An exam point is not theoretically divisible into an infinite number of fractions. However, exam points range over continuous whole numbers and can often be broken into fractions (although not an infinite number) and are therefore often treated as being on a continuous scale. The data in Table 2.3 and Figure 2.1 are reproduced as a frequency polygon in Figure 2.2.

Figure 2.2
A frequency polygon of the data in Table 2.3: class interval size of 3.

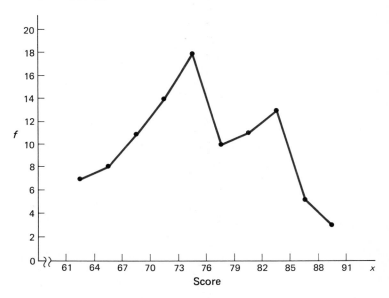

Histograms and frequency polygons are sometimes presented with *relative frequencies* rather than absolute frequencies shown on the *y* axis. *Relative frequencies are determined by forming a fraction for each class interval of the frequency count for that interval divided by the total frequency count for all intervals.* This conversion does not change the shape of the graph, as shown by the relative frequency scale in Figure 2.3.

Shapes of Frequency Distributions

The shapes of frequency distributions, whether they be represented as histograms or frequency polygons, give important information about the sets of scores represented by these graphs. The following sections describe terms used to describe the shape of a frequency distribution.

Figure 2.3
A relative frequency polygon of the data in Figure 2.2.

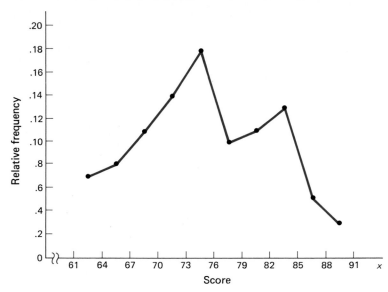

Symmetry   *Symmetry means that one half of something is a mirror image of the other half.* With a symmetrical frequency distribution, each half will look the same from center to outside. In Figure 2.4, graphs *a, b,* and *c* are symmetrical.

skewness:

the extent to which one end of a distribution extends further from the center of the distribution than the other end

Skewness   *By skewness, we mean that one end or ''tail'' of a graph is longer than the other.* In terms of scores, skewness means that there are more scores further from the center (i.e., the highest point in the graph) in one direction than in the other. Skewed right (i.e., skewed positive) means that the tail is longest on the right side. Skewed left (i.e., skewed negative) means that the tail is longest on the left side. In Figure 2.4, graph *d* is skewed right and graph *e* is skewed left. By definition, if a graph is symmetrical, it is not skewed.

To illustrate, try to determine skewness in the following example. There are 50 students in a class, most (but not all) of whom do their reading assignment before every class. The teacher gives an easy (if you have read the assignment) 20-point pop quiz. Will the resulting frequency distribution representing scores on the quiz be symmetrical, skewed right, or skewed left? Let's take a look at the situation. Since it is an easy quiz and since most of the students have read the assignment, most of the quiz scores will be very high, say between 16 and 20. What about the rest of the scores? Likely, some students read some, but not all, of the assignment and got scores in the 5 to 15 range. For those students having better things to do than open their

Figure 2.4
Shapes of frequency distributions.
(*a*) Unimodal symmetric; (*b*) bimodal symmetric; (*c*) unimodal symmetric;
(*d*) *J* curve, skewed right; (*e*) unimodal, skewed left; (*f*) bimodal.

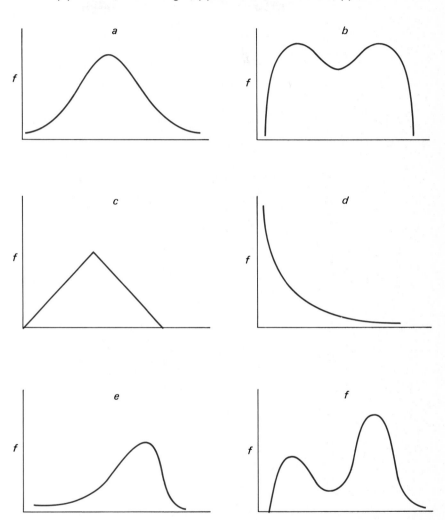

books, their scores ranged from, say, 0 to 5. Typically then, in the situation described, the frequency distribution would be skewed left as illustrated in Figure 2.4*e*, with the *x* axis marked off from 0 to 20 to represent the quiz scores.

Modality *Modality refers to the number of high points or "humps" in a frequency distribution.* In Figure 2.4, *a, c, d,* and *e* are unimodal (i.e., one hump), and *b* and *f* are bimodal (i.e., two humps). It is theoreti-

modality:

the number of high points in a frequency distribution

cally possible to have any number of modes in a frequency distribution, but with real-life situations it is rare to see distributions that have more modes than two. An example of a bimodal distribution is a frequency distribution containing the heights of a large equal number of both male and female college students. The most frequent height for females would be a few inches shorter than for males, so the distributions containing both males and females would look like Figure 2.4*b*.

With respect to sets of scores representing data from experiments in the behavioral sciences, a common frequency distribution is a symmetric, unimodal, bell-shaped distribution called a ***normal distribution*** (Figure 2.4*a*). We will have much more to say about normal distributions in Chapter 5.

SUMMARY

Descriptive statistics allow us to describe a set of numbers in some summarized form. The purpose of *inferential statistics* is to infer characteristics of populations based on samples from that population. A *population* is a group of things, individuals, or measurements that share at least one common characteristic. A *sample* is a subset of a population. A representative sample means that characteristics of the sample closely match that of the population. Two ways of making a sample representation are (a) to *stratify* the sample, that is, define important characteristics of a population and then select samples that have these attributes in the same proportion as does the population, and (b) *random sampling,* that is, selecting by a method such that each member of a population has an equal chance of being chosen for any designation or grouping.

On a statistical basis, one can only generalize the results of a random sample to the population from which that random sample was drawn. Further generalization is based on nonstatistical reasoning about the relationship between the population from which we sampled and a broader population.

A *raw score frequency distribution* is a way of organizing a group of scores such that a frequency count is displayed for the sequentially displayed scores. A *cumulative frequency distribution* cumulates or adds frequencies, starting at the low score and going to the high score. A *percentile* is a number at or below which a certain percentage of ranked scores falls. A *grouped frequency distribution* combines scores into regular class intervals. The size of the class interval is usually set such that about 8–15 intervals are formed.

A *graph* is a visual aid that uses two axes and allows one to more quickly and easily determine certain characteristics of data. Two common types of graphs are *histograms* and *frequency polygons* (line graphs). A histogram is usually used for discrete data, and a frequency polygon is most appropriate for continuous data. The following terms are used to describe the shape of a frequency distribution: (a) *symmetry*—one half being a mirror image of the other half; (b) *skewness*—the direction of the ''tail;'' skewed right (i.e., positive skew) means a long tail on the right, whereas skewed left (i.e., negative skew) means a long tail on the left; (c) *modality*—the number of high points or ''humps'' in a distribution. Most frequency distributions are unimodal (i.e., one hump) or bimodal (i.e., two humps). Data from experiments are often distributed *normally*, meaning that a frequency distribution of these data would be unimodal, symmetric, and bell-shaped.

KEY TERMS

descriptive statistics

inferential statistics

population

sample

representative sample

random sample

parameter

statistic

generalization

raw score frequency distribution

cumulative frequency distribution

grouped frequency distribution

histogram

frequency polygon

symmetry

skewness

modality

discrete scale

continuous scale

percentile

PROBLEMS

1. *a* Suppose you wish to interview 30 randomly selected students from a particular university. How would you randomly select them?

   b How would you randomly place 20 people into four equal-sized groups?

2. How do you know whether a characteristic of a set of data, such as the mean, is a statistic or a parameter?

3. Use a random number table (Table A in Appendix B) to randomly select 40 digits from 0 to 9. Then (a) List them in the order in which they were chosen. (b) Make a raw score frequency distribution from these numbers. (c) Display this distribution as a histogram.

4. *a* Group the 40 scores from problem 3 into 10 groups with 4 scores in each group. Group the scores in the order in which they were chosen. Now find the mean of the 10 groups and make a fre-

quency polygon of these 10 means after rounding each mean to the nearest whole number.

b   Compare the shape of this distribution to that in problem 3.

c   Which is most similar to a normal distribution? (Remember this example. It reflects an important principle that we will discuss in Chapter 5.)

5. Make a raw score frequency distribution from the following set of numbers: 19, 29, 17, 21, 23, 19, 21, 25, 39, 31, 15, 19, 28, 35, 40, 36, 39, 31, 17, 18, 24, 30, 36, 35, 31, 28, 29, 26, 18, 29, 28, 26, 25, 33, 33, 38, 27, 31, 24, 29.

6. a   Convert the raw score frequency distribution in problem 5 to a cumulative frequency distribution.

b   Find the 25th, 50th, and 75th percentiles.

7. a   Draw a frequency polygon from the data in problem 5.

b   Draw a frequency polygon from the data in problem 5, using a relative frequency scale on the *y* axis.

8. Compile a grouped frequency distribution from the data in problem 5. Then draw a frequency polygon from the grouped distribution.

9. Compare the shape of the polygon in problems 8 and 7 in terms of width and variability.

10. Repeat problems 5–9 with the following data set: 7, 9, 1, 20, 24, 19, 3, 6, 5, 11, 5, 9, 3, 7, 5, 2, 21, 19, 22, 7, 15, 13, 6, 4, 19, 9, 7, 3, 5, 14, 22, 20, 3, 21, 12, 8, 4, 11, 19, 21, 15, 6, 9, 10.

11. Give examples of real-world data that you believe would result in the following types of frequency polygons. Justify your example. (Example: normal curve—grades on an exam from many students.) Do *not* use examples from the chapter.

a   unimodal skewed right

b   unimodal skewed left

c   symmetric, not bell-shaped

d   symmetric, bell-shaped

12. Make up 20 scores that would yield a unimodal symmetric distribution and show that this is so by making a histogram of the scores. [*Hint:* Use a limited range and put the most frequent score in the middle of the range. The next most frequent score would be one unit higher and also one unit lower than the most frequent score, etc.]

13. Make up 20 scores that would yield a bimodal symmetric distribution and show it does by making a frequency polygon of the set of numbers.

CHAPTER 3

Central Tendency and Variability

MEASURES OF CENTRAL TENDENCY

A frequency distribution is a convenient way of summarizing and organizing a large set of numbers. However, suppose our primary concern is not with presenting all the data in some form whereby the spread and midrange of scores are shown. Suppose our concern is with presenting a single number that typifies the whole set of numbers. What number would best do this? Certainly, the number should be from somewhere in the middle of the range of scores in the

central tendency:

the middle of a group; typical measures of central tendency are the mean, median, and mode

set. *Measures that reflect this "middle" score, this "typical" score, this "average" score are called **measures of central tendency.*** There are three common measures of central tendency: the *mean*, the *median*, and the *mode*. Each measure is computed differently, and each will usually, but not always, yield a different value.

THE MEAN

Computation

The mean is so common that you likely already know what it means and how to compute it. However, it will set a useful precedent to state how the mean is computed in algebraic form. The symbols are

$$\bar{X} = \frac{\Sigma X}{N}$$

where $\bar{X}$ = mean of a set of scores
ΣX = sum of all of the scores in the set
N = total number of scores in the set

A convenient shorthand used in most statistics books and throughout the rest of this book is to let a letter, such as X, stand for all of the scores (e.g., numbers) in a set. A bar over a letter (e.g., $\bar{X}$) represents the mean for the set of numbers that the letter represents. The symbol Σ (sigma) is a summation sign and always means "add all of the numbers in the set represented by the symbol or symbols that follow the Σ." Thus ΣX means "sum all of the numbers represented by X."

To illustrate, suppose we have a set of seven quiz scores: 5, 7, 10, 5, 8, 8, 6. Let X represent those seven scores; ΣX is then 5 + 7 + 10 + 5 + 8 + 8 + 6 = 49, and N = 7. Therefore

$$\bar{X} = \frac{\Sigma X}{N} = \frac{49}{7} = 7.0$$

Properties of the Mean

The mean has an important mathematical property that is often a useful computational check: ***the sum of the deviations from the mean is zero.*** Symbolically, this is written

$$\Sigma (X - \bar{X}) = 0$$

Remember, X stands for each score in a set of scores, and $\bar{X}$ is the mean of the set. If the set of scores is 4, 5, 7, 10, 12, 22, the mean $\bar{X}$ is

$$\bar{X} = \frac{\Sigma X}{N} = \frac{4+5+7+10+12+22}{6} = \frac{60}{6} = 10$$

Then

$$\Sigma (X - \bar{X}) = (4-10)+(5-10)+(7-10)+(10-10)+(12-10)+(22-10)$$

$$= -6-5-3+0+2+12 = 0$$

The following is another set of six scores where $\bar{X} = 10$. We will align them in a column along with the first set of scores in Table 3.1.

TABLE 3.1
Illustrating $\Sigma(X - \bar{X})$ for Two Sets of Scores

Score (X)	$X - \bar{X}$	Score (X)	$X - \bar{X}$
22	+12	13	+3
12	+2	11	+1
10	0	11	+1
7	−3	9	−1
5	−5	8	−2
4	−6	8	−2
$\Sigma X = 60$	$\Sigma(X - \bar{X}) = 0$	$\Sigma X = 60$	$\Sigma(X - \bar{X}) = 0$
$\bar{X} = 10$		$\bar{X} = 10$	

As always, in both sets of scores $\Sigma(X - \bar{X}) = 0$. Now, notice which set of scores has the widest score spread and then notice how that is reflected in the $X - \bar{X}$ column. The absolute values of the numbers in the $X - \bar{X}$ column for the first set are much larger, reflecting the increased score spread. This will be an important principle when we shortly discuss the topic of variability.

A second property of the mean is that *a change in any score in a set will always change the mean of the set.* This is a unique property of the mean, compared with the other measures of central tendency.

An important characteristic of the mean is that the best estimate of an unknown population mean is the mean of a random sample from that population. We will not expand on the importance of this principle here, but we will expand on it in Chapter 5 and in Chapter 12, when we talk about parameter estimates.

THE MEDIAN

median:

measure of central tendency that is the center score of a ranked series of scores

*The **median** is the number that divides a ranked set of scores such that half of the scores are above this number and half of the scores are below it.* Computing the median, therefore, involves ranking the scores from high to low and then determining the middle score. There are two situations involving the median: the simple one in which the middle score occurs only once, and the complex one in which the middle score occurs more than once. In this book, we will deal only with the simple, since we are interested more in the concept than in broad applications. In fact, the median is rarely used in research where inferences are drawn from samples to populations.

In the simple case: (a) If you have an odd number of scores, and the middle score occurs only once, then the median is that middle score. For example:

21
20
20
15 = median
9
5
1

The median is 15. (b) If you have an even number of scores, and no repeated scores that are the same as the median, then the median is interpolated to be the midpoint between the two middle scores. For example:

34
31
30
26   24.5 = median
23
20
19
7

The median is interpreted to be halfway between 26 and 23 or 24.5.

THE MODE

mode:

measure of central tendency that is the most frequent score in a group of scores

*The **mode** is simply the score that occurs most frequently within a set of scores.* If two scores are tied for the most frequent score, then both scores are considered to be modes for that set. When a set of scores is displayed as a histogram or frequency polygon, the mode (i.e., the most frequent score) will be the score corresponding to the high point in the figure. Notice that some liberties were taken with the mode when we talked about modality as a way to describe shapes of frequency distributions. We called the graph of Figure 2.3*f* bimodal even though the frequency of the smaller ''hump'' is not as large as the larger hump. Figure 2.3*b* is truly bimodal. This potential confusion is not as serious as it might seem, since the mode is not a good measure of central tendency, and therefore rarely used by itself as *the* measure of central tendency.

MEASURES OF CENTRAL TENDENCY AND SHAPES OF FREQUENCY DISTRIBUTIONS

Figure 3.1 shows the relationship between shapes of frequency distributions and the mean, median, and mode for unimodal distributions. In unimodal symmetric distributions, the three are always the same. In unimodal skewed distributions, the mean will always be the farthest of the three in the direction of the tail. This is because, as we will shortly demonstrate, the mean is influenced by extreme scores, whereas the median is not and the mode is usually not.

Figure 3.1

Central tendency and shapes of frequency distribution: (*a*) symmetrical; (*b*) skewed positive; (*c*) skewed negative.

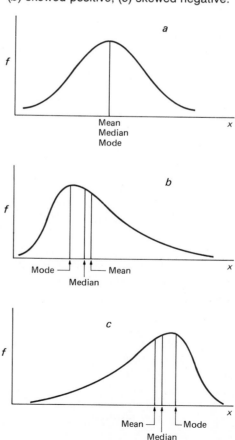

WHICH MEASURE OF CENTRAL TENDENCY SHOULD YOU USE?

Since the mode can occur near the end points of a frequency distribution, it is rarely used as *the* measure of central tendency. For example, suppose a teacher gives a quiz in which most students do pretty well except for the five students who got zeros because they cut the class. The mode of this quiz might well be zero, and zero clearly would not be adequate as a typical or representative score.

The choice between the mean and median depends on whether one wants a measure of central tendency that reflects extreme scores that occur only at one end of a distribution of scores. The following set of exam scores will illustrate the situation:

Scores	
97	
93	
90	
87	
86	
84	median = 84
81	
81	
80	$\bar{X} = 73.7$
23	
09	

In choosing a measure of central tendency, the class would obviously vote for the mean, thinking that their score will therefore look better since the mean is smaller than the median. However, the teacher may not see it that way. He might think that the median is a more representative score. The choice between using a mean or a median is made by the user of the data. When presenting a data summary to others, often both measures are given so that the user can make his or her own choice.

In contrast to the mean, notice that the lowering of any score that is already below the median by any amount will not affect the median. Also raising any score that is already above the median by any amount will not affect the median.

With experiments, the situation is somewhat different than with exam scores. Here, the experimenter usually wants to include all scores from a sample and to have each score influence the measure of central tendency. This suggests a preference for the mean. Also, the mean of a sample is a more accurate estimate of a population mean (the usual parameter of interest in an experiment) than is a sample median. Further, most statistical tests of inference test for differences between means, not medians. For all of these reasons, the mean is almost always used as the measure by which comparisons are made between the independent variable conditions in an experiment.

MEASURES OF VARIABILITY

When we talked about central tendency, we discussed measures that would yield a single number to represent a typical or average score from a set of scores. However, no measure of central tendency says anything at all about the scores in terms of their being close together or spread apart. *Variability means score spread,* and measures that reflect this are very useful tools in summarizing a set of data. Before discussing specific ways of representing variability within a set of scores, we will discuss variability in general and why it is so important in research in the social sciences.

Variability occurs both within and between individuals. When we say that a sports team is consistent, we mean that its performance has low variability. The team is good, bad, or mediocre, but not all three. Variability also relates to probability. A team that is consistently good has a high probability of playing a good game the next time it plays. Likewise, the consistently bad team will likely play a bad game next time. If we say that there is a lot of variability between the students in a class, we mean that the students range from very good to very poor. The same, of course, is true for teachers within an institution. High variability means some very good teachers and some very poor, whereas low variability means similar in quality, be it good or bad. In all living organisms, variability on all traits within members of a species is the rule. In fact, this principle was a central postulate in Darwin's original theory of evolution and remains so today. Thus, any random selection of people will vary on things like height, weight, opinions, knowledge-based questions, or in performance on about any type of test. The degree of variability is sometimes of utmost importance. For example, during the Vietnam War, there was great variability in support of the war. This led to large numbers of people in opposite camps, creating a strong division within the country.

Dealing with high variability across individuals is a fact of life for the social scientist. Complicating the matter further is the fact that people also vary within themselves across time. An athlete with high variability is inconsistent, doing very well sometimes and very poorly other times. A similar analogy can be made for artists, students, teachers, TV personalities, and so on. It is because of the high variability across individuals, and sometimes within individuals, that statistical procedures are so important to the social scientist. As we will see in later chapters, **statistics are the window that the social scientist must look through to find the order necessary for scientific progress.**

THE RANGE

range:

a measure of variability represented by the highest number in a set minus the lowest number in the set

*The **range** is simply the difference between the high and low scores within a set of scores.* The range is to variability as the mode is to central tendency—quick, easy, and not very stable. The largest drawback with using the range as *the* measure of variability is that it uses only two scores in its computation, the top and bottom. Thus, all of the other scores could be highly bunched or spread out along the whole range of scores. One simply does not know this when only the range is given. A better way of representing variability would be to develop a technique whereby every score within a set would contribute to the measure. The most common measure of variability, the *standard deviation*, does this.

THE STANDARD DEVIATION

standard deviation:

a measure of score spread or variability

As the mean was a measure of central tendency defined by a formula (i.e., $\Sigma X/N$), so is the standard deviation a measure of variability defined by a formula. Although the formula is more complicated for the standard deviation than for the mean, you need to fully understand it in order to fully understand the meaning of the standard deviation. We will first discuss some background for the standard deviation, and then we will present the specific formula for its development.

There are several possible ways that one might measure score spread or variability. For example, one might rank the scores from high to low and take the mean distance between each adjacent pair of scores as a measure of variability. Or, one could first find the mean and then subtract each score from the mean while ignoring the sign of the scores (see Table 3.1). The mean of these scores (i.e., $\Sigma |X - \bar{X}|/N$ is called the *average deviation* and is also a measure of variability. This technique as well as the standard deviation is an adequate way of measuring score spread. The advantage that the standard deviation has is that it is related to the normal curve and the development of inferential statistics, whereas the average deviation is not. Thus the standard deviation has become the most common measure of variability.

Population Standard Deviation

The standard deviation of a population of scores is

$$\sigma = \sqrt{\Sigma \frac{(X - \bar{X})^2}{N}}$$

You can best see what is happening if we go through an example step by step. Memorizing these steps will be very helpful in retaining the meaning of the standard deviation. The following four steps will yield the standard deviation for a population of scores.

Step 1. Find $\bar{X}$: $\bar{X} = \Sigma X/N$.

Add the scores and divide by the total number of scores.

Step 2. Find SS (sum of squared deviations): $SS = \Sigma (X - \bar{X})^2$.

Take each score and subtract the mean from it, square each answer (called a deviation score), and add all of the squared deviation scores.

Step 3. Find the mean of SS, that is, SS/N.

Take the SS value and divide by the total number of scores.

Step 4. The standard deviation (σ) is $\sqrt{SS/N}$

Take the square root of the SS/N value.

Example 1. Let X represent the following set of scores: 9, 13, 12, 8, 8.

Step 1.

$$\bar{X} = \frac{9 + 13 + 12 + 8 + 8}{5} = \frac{50}{5} = 10$$

Step 2.

$$SS = \Sigma (X - \bar{X})^2 = (9 - 10)^2 + (13 - 10)^2 + (12 - 10)^2 +$$

$$(8-10)^2 + (8-10)^2 = 1^2 + 3^2 + 2^2 + 2^2 + 2^2 = 1+9+4+4+4 = 22$$

It is usually more convenient to do steps 1, 2, and 3 in columns as follows:

X	$X - \bar{X}$	$(X - \bar{X})^2$
9	-1	1
13	$+3$	9
12	$+2$	4
8	-2	4
8	-2	4
$\Sigma X = 50$	$\Sigma (X - \bar{X}) = 0$	$22 = \Sigma (X - \bar{X})^2$

Repeating steps 1 and 2 gives

Step 1.

$$\bar{X} = \frac{\Sigma X}{N} = \frac{50}{5} = 10$$

Step 2.

$$SS = \Sigma\,(X - \bar{X})^2 = 22$$

Step 3.

$$\frac{SS}{N} = \frac{22}{5} = 4.40$$

Step 4.

$$\sigma = \sqrt{\frac{SS}{N}} = \sqrt{\frac{\Sigma\,(X - \bar{X})^2}{N}} = \sqrt{4.40} = 2.10$$

Example 2. (in column format)
Find σ for the following scores: 12, 9, 0, 3, 5, 2, 7, 10.

X	$X - \bar{X}$	$(X - \bar{X})^2$
12	+6	36
9	+3	9
0	−6	36
3	−3	9
5	−1	1
2	−4	16
7	+1	1
10	+4	16
$\Sigma X = 48$	$\Sigma\,(X - \bar{X}) = 0$	$124 = \Sigma\,(X - \bar{X})^2 = SS\,(\text{step2})$

Step 1.

$$\bar{X} = \frac{48}{8} = 6$$

Step 3.

$$\Sigma\,\frac{(X - \bar{X})^2}{N} = \frac{SS}{N} = \frac{124}{8} = 15.5$$

Step 4.

$$\sqrt{\frac{SS}{N}} = \sqrt{\frac{\Sigma\,(X - \bar{X})^2}{N}} = \sqrt{15.50} = 3.94$$

From these examples, we can see that putting all four steps into one algebraic formula gives

$$\sigma = \sqrt{\frac{\Sigma\,(X - \bar{X})^2}{N}} \qquad \text{3.1}$$

Some Properties of the Standard Deviation

(a)   Unless scores are all the same (in which case the range and standard deviation are zero), the standard deviation will always be less than the range. An estimate of one third of the range will usually be close for nonnormal distributions.

(b)   The standard deviation will always be either zero (only if all scores are the same) or positive (because all deviations are squared).

(c)   In normal distributions, about two thirds of the scores will be within 1 standard deviation of the mean and the standard deviation will be about one sixth of the range.

These properties are useful "eyeball" checks on your calculation of standard deviations. Let us illustrate these principles by estimating the σ of four sets of scores. Write down the alternative that you think is correct for each.

A	B	C	D
4	34	99	39
6	21	102	39
9	30	98	39
−3	26	100	
11	39	97	
−6	40	103	
1	32	101	

A	B	C	D
(a) = 19	(a) = 0	(a) = 6	(a) = 0
(b) = −6	(b) = 2	(b) = 2	(b) = 10
(c) = +7	(c) = 6	(c) = 99	(c) = 20

Would (a), (b), or (c) be the best guess for the σ of A? The range of scores is from −6 to 11, which equals 17. Therefore, alternative (a) cannot be correct. Alternative (b) cannot be correct since σ can never be negative. Alternative (c) is a good guess since it is not far from one third of the range. The best guess for the σ of B is alternative (c), for problem C the answer is (b), and the σ for D is zero.

Estimating Standard Deviations from Samples

Suppose that the total number of scores from a population is not available for some reason, so we have to estimate σ from a random sample from that population. If we computed a standard deviation from a random sample in the same way that we have been doing for a population (i.e.,

$$\sigma = \sqrt{\frac{\Sigma (X - \bar{X})^2}{N}}$$

would that be the best estimate that we could make of the population standard deviation? Statisticians have determined that the answer is almost, but not quite. The best estimate is given if we substitute $N - 1$

for N in the formula for σ. The *estimated* population standard deviation from a random sample (symbolized by S) is then

$$S = \sqrt{\frac{\Sigma\,(X - \bar{X})^2}{N - 1}} \qquad\qquad 3.2$$

The steps and procedures for S are identical to those for σ, with N replaced by $N-1$. For large N, the values for S and σ are very similar.

How Do I Know If a Set of Scores Is a Sample or a Population?

Remember, a population contains the total number of scores or potential scores. A sample contains only part of the set of scores. Therefore, to determine if a set of scores is a sample or population, one must determine the context in which the scores are to be used. If the interest in a set of scores does not go beyond that specific set of scores, then the set is a population. If the interest in the set of scores is extended beyond that set, then the set is a sample. For example, suppose I give my students a test covering a chapter in a text that I have assigned. The test has been sent to me by the book publisher who is also interested in the results of the test since they may wish to include the test as part of an instructor's manual. Do the scores from the test constitute a population or sample? It depends. For me, the test scores are a population. I have no wish to infer anything about a larger population. The students are my population. In computing a standard deviation for these scores, I would therefore use the formula for σ. However, the book publisher wishes to use these scores to estimate how students nationwide would perform on the test. Therefore, the publisher should use the formula for S in computing the standard deviation for this set of scores. Whether σ or S is appropriate is a decision that must be made by the user of the data. Since the interest in conditions in experiments (e.g., the drug condition or the placebo condition) is almost always involved with larger groups than actually participate in the study, the data in experiments are almost always samples. This is why standard deviations from research studies generally use S instead of σ. Estimating a population mean from a sample mean is much simpler than estimating population standard deviations from sample standard deviations. *The best estimate of a population mean (when it is unknown) is the mean of a random sample from that population,* with no corrections in the computation of the sample mean being necessary.

Computational Formula

Question: What is the fastest way of obtaining S or σ from a large set of data?

Answer: Buy yourself a statistics calculator with an S or σ key, enter the numbers, and press the key. Happily for users of statistics, computers and calculators have taken much of the number-crunching drudgery out of the computational process. However, situations remain where it might be necessary (or perhaps less expensive) to use a regular hand calculator to find σ or S. If so, there are algebraically equivalent formulas to the ones we have used to define σ and S. These algebraically equivalent formulas are called ***computing formulas,*** since usually computations will be faster by using them. The formulas that we have thus far used are called ***definitional formulas.*** The problem with using a definitional formula to compute σ or S is that, if $\overline{X}$ is not a whole number, the term $X - \overline{X}$ will contain fractions. If we convert the fractions into decimals and square these numbers, a lot of labor is involved and a small amount of precision may be lost because of rounding numbers. It is therefore desirable from a computational perspective to substitute for the SS expression, $\Sigma\,(X - \overline{X})^2$, by using the algebraically equivalent expression $(N\Sigma\,X^2 - (\Sigma\,X)^2)/N$.

The computational formulas for σ and S are then

$$\sigma = \sqrt{\frac{N\Sigma\,X^2 = (\Sigma\,X)^2}{N^2}} \qquad\qquad 3.3$$

and

$$S = \sqrt{\frac{N\Sigma\,X^2 - (\Sigma\,X)^2}{N(N-1)}} \qquad\qquad 3.4$$

Important. **Note that $\Sigma\,X^2$ means to add up all of the X scores after squaring them. The expression $(\Sigma X)^2$ says to add the X scores together, then square the sum.**

$\Sigma\,X^2$ is not the same as $(\Sigma\,X)^2$

Procedures for finding S and σ using both the definitional and computational formulas are shown in Box 3.1.

THE VARIANCE

Variance:

a measure of score spread defined as the square of the standard deviation

Another measure of variability is called the *Variance*. The **Variance** is *simply the square of the standard deviation* and is symbolically expressed as S^2 or σ^2, depending on whether the referent is a sample or a population. In our four-step procedure for finding a standard deviation, the Variance is the term obtained after step 3, before the square root in step 4 is obtained. To prevent confusion between *variability* (score spread) and *Variance* (σ^2 or Σ^2), we will always capitalize the Variance meaning S^2 or σ^2. The Variance is seldom used as a descriptive statistic, but is often found in formulas involving inferential statistics.

BOX 3.1

Computing σ and S from Definitional and Computational Formulas

Definitional Formula			Computational Formula	
X	$X - \bar{X}$	$(X - \bar{X})^2$	X	X^2
0	-5	25	0	0
7	$+2$	4	7	49
1	-4	16	1	1
6	$+1$	1	6	36
8	$+3$	9	8	64
8	$+3$	9	8	64

$\Sigma X = 30$

$N = 6$ $64 = \Sigma(X - \bar{X})^2$ $30 = \Sigma X$ $214 = \Sigma X^2$

                                                      $N = 6$

$$\bar{X} = \frac{\Sigma X}{N} = \frac{30}{6} = 5$$

$$\sigma = \sqrt{\frac{\Sigma(X - \bar{X})^2}{N}} = \sqrt{\frac{64}{6}}$$

$$= \sqrt{10.67} = 3.27$$

$$S = \sqrt{\frac{\Sigma(X - \bar{X})^2}{N - 1}} = \sqrt{\frac{64}{5}}$$

$$= \sqrt{12.8} = 3.58$$

$$\sigma = \sqrt{\frac{N\Sigma X^2 - (\Sigma X)^2}{N^2}} = \sqrt{\frac{6(214) - (30)^2}{(6)(6)}}$$

$$= \sqrt{10.67} = 3.27$$

$$S = \sqrt{\frac{N\Sigma X^2 - (\Sigma X)^2}{N(N - 1)}} = \sqrt{\frac{6(214) - (30)^2}{(6)(5)}}$$

$$= \sqrt{12.8} = 3.58$$

	BOX 3.2 Summary of Symbols and Formulas		
Symbol	Definition	Definitional Formula	Computational Formula
μ	Population mean	$\Sigma X/N$	$\Sigma X/N$
$\bar{X}$	Sample mean used to estimate population mean	$\Sigma X/N$	$\Sigma X/N$
SS	Sum of squared deviations	$\Sigma(X - \bar{X})^2$	$(N\Sigma X^2 - (\Sigma X)^2)/N$
σ	Population standard deviation	$\sqrt{SS/N}$	$\sqrt{SS/N}$
S	Sample standard deviation used to estimate σ	$\sqrt{SS/(N-1)}$	$\sqrt{SS/(N-1)}$
σ^2	Population Variance	SS/N	SS/N
S^2	Sample Variance used to estimate σ^2	$SS/(N-1)$	$SS/(N-1)$

SUMMARY

A *measure of central tendency* is a number that represents a typical or middle score from a set of scores. The most common measure of central tendency is the *mean*. The mean is computed by taking the sum of all of the scores in a set and dividing by the number of scores in that set. One property of the mean is that the sum of each score in a set minus the mean will always be zero ($\Sigma (X - \bar{X}) = 0$). A second property of the mean is that a change in any score within the set of scores will change the mean of the set.

A second measure of central tendency is the *median*. The median is defined as the number that divides a ranked distribution such that half of the scores are above this number and half are below it. When there is an even number of scores in a set, the median is interpolated to be the midpoint between the two center scores. A third measure of central tendency is the *mode*, defined as the score that occurs most frequently within a set.

In unimodal symmetric distributions, the mean, median, and mode have the same value. In unimodal skewed distributions, the mean will always be the farthest of the three in the direction of the tail.

The mode is rarely used as *the* measure of central tendency because it can occur at or near the end of a distribution. The choice between the median and the mean depends on the decision that the user of the data makes about extreme scores that occur only in one end of a distribution of scores. If the user wants a measure of central tendency that reflects extreme scores, he or she chooses the mean; if the effect of extreme scores is to be disregarded, the median is the most appropriate measure of central tendency.

With respect to a set of a data, *variability* means score spread. The simplest measure of variability is the *range.* The range is the difference between the high score and the low score within a set of scores. The most common measure of variability is the *standard deviation.* The standard deviation of a population (σ) is defined by the following formulas:

$$\sigma = \sqrt{\frac{\Sigma\,(X - \bar{X})^2}{N}} \quad \text{or} \quad \sigma = \sqrt{\frac{SS}{N}}$$

Some properties of the standard deviation are (a) when the scores in a set are not the same, σ will always be less than the range; (b) σ will always be either zero (only if all scores in a set are the same) or positive; (c) in normal distributions, about two thirds of the scores will be within 1 standard deviation of the mean.

When a standard deviation for a population has to be estimated from a sample, $\Sigma\,(X - \bar{X})^2$ is divided by $N - 1$ instead of N to yield the formula

$$S = \sqrt{\frac{\Sigma\,(X - \bar{X})^2}{N - 1}} \quad \text{or} \quad S = \sqrt{\frac{SS}{N - 1}}$$

One determines whether a set of scores is a sample or a population by examining the context in which the scores are to be used. If the interest in the scores extends beyond the set, then the set is a sample. If the interest does not go beyond the set, then the set is a population.

Computing formulas are algebraically equivalent to definitional formulas and are usually faster to work with, particularly in situations where the mean is not a whole number.

There is a measure of variability called the *Variance.* It is simply the square of the standard deviation. The Variance is seldom used as a descriptive statistic, but is often found in formulas involving inferential statistics.

KEY TERMS

mean	σ^2
median	σ

mode S^2
variability S
central tendency   SS
$\overline{X}$ standard deviation
μ range
                              Variance

PROBLEMS

1. Find the mean, median, and mode for the following set of numbers: 27, 25, 25, 19, 15, 10, 10, 10, 9, 8, 7, 7, 5, 3, 2, 1, 1.

2. Find the mean, median, and mode for the following set of numbers: 6, −5, 9, 7, 0, −3, 7, 0.

3. Add a fourth number to the following three numbers so that the mean of the four numbers is 10: 15, 11, 7.

4. Make up a set of six different numbers that have a mean of 15 and show that the sum of the deviation scores is zero (i.e., $\Sigma(X - \overline{X}) = 0$).

5. Suppose a set of scores has a mean of 50. If the score, 20, is added to the set, another score of _____ must be added if the mean of 50 is to be maintained. Complete the following general statement. If a number below the mean is added to a set of scores, then _____ if the mean of the set is to remain the same.

6. There are five Smith brothers. Their mean income is $250 per week, and their median income is $180 per week. Joe, the lowest paid, gets fired from his $100 a week job and now has an income of $0 per week.

    a What is the new mean weekly income of the five brothers?

    b What is the new median weekly income of the five brothers?

7. Is the mean or median the most appropriate in the following situations? Justify your answer.

    a Describing incomes in a district of a city that has a few very wealthy people.

    b Describing SAT scores for incoming freshmen in a large university.

    c Describing exam scores where five people get zeros because of absence.

    d You are applying for a grant based on need and must report your "average" income for the last five years. Your incomes were $10,000, $11,000, $12,000, $0, $8,000.

8. Draw a frequency polygon representing your estimate of the number of workers earning various hourly wages for the following companies. The x axis will be amount of hourly wage, the y axis will be an estimated frequency count. Mark where you think the mean and median would be within the graph.

　　a　General Motors

　　b　McDonald's (a high percentage of minimum-wage employees)

9. Use the definitional formulas to find σ and S for the following set of scores: 3, 1, 8, 9, 5, 4.

10. Add 10 to each of the scores in problem 9 and find σ (use the definitional formula). State in your own words why adding or subtracting a constant (any number) to all members of a set of numbers will not change the standard deviation.

11. For the following scores, estimate S and then use the definitional formula to compute it.

a	b	c	d
21	105	35	38
19	99	17	35
30	94	21	26
10	101	28	18
15	100	31	24
25	108	19	30
20	103	14	17
	88	33	29

12. Use the computational formula to verify the answers in problem 11. You may get small differences because of rounding.

13. In reporting variability on exams to students, should an instructor use σ or S? Explain.

14. Find σ for the following set of scores: 5, 7, 9, 4, 11, 1, 13.

15. Find σ for the scores in problem 14 after multiplying each number by 5.

16. How many times larger is the σ in problem 15 than the σ in problem 14? (The formula is $\sigma_{new} = k\sigma_{old}$.)

17. Upon finding an S value of -7.3, what should you immediately know?

18. The following scores represent the performance in hundredths of a second of 10 people taking a simple reaction time test where a key is depressed when a person sees a light flash. Find $\overline{X}$ and S: 28, 32, 30, 34, 29, 30, 30, 29, 33, 31.

19. Suppose the task in problem 18 was done in very noisy and chaotic surroundings that caused a longer reaction time, ranging from 1 to 9 hundredths of a second. Since the noise varied greatly, the length of added reaction time, ranging from 1 to 9 hundredths of a second, varied randomly between people. From the random number table (Table A in the appendix), add a random digit from 1 to 9 to each of the 10 scores in problem 18 to represent the effects of the noise. Compute the new $\bar{X}$ and S and compare in magnitude to the old $\bar{X}$ and S.

20. Suppose you are trying to estimate a population mean from a sample mean, and there is a lot of variability in the scores from which you compute the sample mean. How will this high variability affect the accuracy of predicting the population mean? Why?

CHAPTER 4

Statistical Inference

THE ROLE OF INFERENTIAL STATISTICS WITHIN EXPERIMENTATION

As we saw in Chapter 1, the process of experimentation is very similar to that of solving a crime in a detective story (see Box 4.1). As detectives, we try to eliminate suspects while building a case against the prime suspect. As experimenters, we build our case for the independent variable causing something to happen through an experimental design, but before we can hope to make a case for this we need to eliminate a suspect that is forever hounding experimentalists. This villain is called *chance.* **In any situation where variability is involved, differences between conditions in an experimental situation can occur because of entirely random (i.e., chance) influences on the data.** For example, if one were testing the effects of a drug on reaction time and randomly assigning people to the drug and no-drug groups, it is possible, although unlikely, that most of the fast people would be in one group whereas most of the slow people would be in the other. If this circumstance were true, then some or all of the difference between groups with respect to reaction time would simply be due to the initial difference between the groups and would have nothing to do with the effects of the drug. Thus, if the results of an experiment of this type showed that the people in the drug condition had generally slower reaction times, we would need to demonstrate that the probability of our results coming from a lopsided (with respect to reaction time) arrangement of people is very small before inferring that it was the drug that was causing the slower reaction time.

With respect to determining whether an experimental manipulation has an effect, *the purpose of all statistical tests of inference is to determine the probability that differences in data resulting from the experimental manipulation are entirely due to random influences.* The experimenter must then decide if that probability value is small enough to reject chance (i.e., random influences) as the explanation for the outcome of a set of results. However, usually this decision is not arbitrary; rather, the experimenter uses a conventional decision criterion of 1 chance in 20 or .05. In other words, *if differences in data representing different conditions in an experiment represent a chance occurrence of less than 1 in 20, the experimenter will reject the explanation that the results happened because of some unlikely random occurrence.* When the random influence explanation is rejected, the experimenter can infer, based on the experimental design, that the results of the experiment occurred because of the independent variable manipulation.

BOX 4.1

Alternatives for Experimental Effects

Differences in data between conditions in an experiment can occur because of any or all of the following reasons: (a) random factors (this is the "by chance" explanation for a set of results); (b) uncontrolled or extraneous variables; (c) experimental manipulation.

The purpose of inferential statistics is to assess the first alternative and eliminate it if it is unlikely.

The purpose of experimental designs is to eliminate the second alternative while supporting the third.

TECHNIQUES FOR DETERMINING PROBABILITY VALUES

There are two ways of determining the probability that a set of experimental results has been obtained through random influences. One way utilizes what are called *nonparametric techniques,* in which prob-

nonparametric statistical inference:

inference techniques that do not involve theoretical models or distributions, but are based only on random sequences of events

parametric statistical inference:

inference techniques that involve assumptions about population parameters and use theoretical distributions

ability values are often computed from random sequences of events, as in coin-toss situations. For example, we can determine the probability of getting five heads in five coin tosses, as explained later in this chapter. Nonparametric techniques do not involve estimates of parameters, whereas parametric techniques do. Also, with *parametric techniques,* assumptions about the shape of population distributions are made (e.g., that a population is normally distributed). Nonparametric techniques do not involve assumptions about populations. Starting with Chapter 5, most of the tests that we study are parametric techniques. However, the sign test, used in this chapter, is a nonparametric test. After you have studied Chapters 4, 5, and 6, the distinction between parametric and nonparametric tests will be much clearer. Remember, the purpose of both parametric and nonparametric tests is the same—to determine probability values of random influences on data; from this probability value we can decide whether to reject the differences that we see in data as being due to random influences.

ILLUSTRATING STATISTICAL INFERENCE CONCEPTS

Example 1. Let us take an example that might occur in everyday life to illustrate how probability and decision making might work in a situation of uncertainty. Suppose we pay $1 for a raffle ticket that offers a $600 first prize. After the drawing, we find out that the winner was also the sponsor of the raffle. Should we conclude that the raffle was rigged?

We will look at two scenarios and then develop a procedure for making a decision. First, assume that 1000 tickets were sold and that the sponsor bought one. With an honest draw, he or she could win, but the probability is very small—1 chance in 1000 to be exact. Even though we cannot be 100% sure (only 99.9%), the probability of the sponsor winning with an honest draw is so small that we likely would conclude that the draw was dishonest.

Now suppose that only 500 tickets had been sold, so the sponsor, being a good sport, bought the remaining 500 tickets. (We will assume that the $400 profit from the raffle was going to charity.) Now the situation has changed considerably. With an honest draw the sponsor has 500 chances in 1000 or a 50-50 chance to win. If the sponsor won under these conditions, we would likely be much less suspicious of dishonesty than if the sponsor had only bought one ticket. But why? After all, it takes only one ticket to win. The answer lies in understanding probability. If every ticket is equally likely to win, then a person who has 500 tickets is 500 times as likely to win as a person who has one ticket. However, even with a 50-50 chance to win you might still think the sponsor was cheating because the sponsor may have been in a position to rig the draw. In other words, how you interpret the outcome of a probabilistic occurrence involves more than a simple decision based on probabilities. Let us establish steps that would be useful in making a decision in this situation.

Step 1. Based on your analysis of the situation, what probability values representing what outcomes would you be willing to accept as having happened by chance (i.e., what is your criterion for a chance outcome)? You might be dogmatic and say that the sponsor is innocent even if he or she purchased only one ticket and won, or you might say that he or she is guilty even if the sponsor had purchased 900 tickets. But with thought you might decide to go with an outcome somewhere in between.

Step 2. Determine the probability that an outcome has happened by chance (e.g., what is the probability of the sponsor winning, given the number of tickets that he or she bought).

Step 3. Compare the probability outcome in step 2 with your criterion. If the probability value for the outcome is greater than your criterion, you would assume that the outcome is reasonably likely to have happened by chance, and you accept the occurrence as having happened by chance. If the chance level for the outcome in step 2 (i.e., the sponsor winning) is less than your criterion, you would consider the "by chance" explanation to be unreasonable, dismiss it, and say something else caused the outcome. In the present example, the alternative to the "by chance" explanation is that the sponsor cheated.

Example 2. We will now look at a more complicated situation and see how the same three steps in making a decision would apply. Suppose you often have a drink between classes with a friend. For a touch of excitement, your friend suggests flipping a coin to see who pays. The friend takes out a coin, flips it in the air, and slaps it on her wrist. You call it, and she shows you the coin. The problem is that of the 11 times that she has done this, you have only won once. Has your friend been cheating you? Everything that we said about the raffle applies here. Yes, it is possible for you to lose 10 out of 11 coin tosses under honest conditions. No, it is not a likely event to happen under honest conditions. But how unlikely is it? You can get a feel for this probability by doing a simulation. Take 11 coins, shake them, and let them fall on a hard surface and see how many times you have to do this to get at least 10 heads (it likely will take you over 100 drops). However, the probability of getting at least 10 heads in 11 tosses can be computed exactly. You would then treat this probability as step 2 in the three-step decision-making process.

Let's now disgress into probability theory so that you may learn how to compute probabilities that occur in situations such as tossing coins. We will then come back to this example and make a decision about our friend cheating. We will also see how these principles can be applied to making a decision about what happened in a real experiment.

PROBABILITY

Probability can be defined as the likelihood of an event occurring. Probability is measured on a scale ranging from 0 (it could not happen) to 1 (it definitely will happen). There are two ways of determining probabilities. The first is to compute the probability of an event based on the *relative frequency* with which that event has occurred in the past. For example, if 80% of your college grades have been B's and C's, you would expect about the same relative frequency of B's and C's again (assuming conditions affecting grades, such as study time or type of course, do not change). Let's take another example illustrating relative frequency. Suppose when playing *Monopoly*, you observe that you land on a railroad once for about every two trips around (passing GO) the board. You might then conclude that the probability of you (or someone else) going around the board without landing on a railroad is about 1 chance in 2 or 50-50. You might then use this probability estimate to decide on your cash reserve if someone else owned all the railroads.

a priori:

by reason alone

The other way of computing probabilities is called the *a priori* (i.e., by reason alone) method. With this method probabilities are determined based on knowledge of an event independent of relative frequencies. For example, we can determine the probability of rolling a die and getting a 6 based on facts about the die. If it has an equal surface on each side with the weight also equally distributed, we can determine that the probability of rolling the die and getting a 6 is 1 in 6. Using the same logic, we can determine that the probability of tossing a coin and getting a head is 1 in 2, 50-50, or 50%, all three probability statements meaning the same thing. For now we will only be concerned with probabilities that can be determined *a priori.*

For our purposes, we will limit the discussion of probability to discrete events such as coin tosses, where outcomes will be defined. We will call the total range of outcomes a *set,* and it will represent a probability of 1. For example, the set of outcomes for a coin toss would be head (*H*) or tail (*T*); that is, two outcomes. So the probability of getting either an *H* or a *T* exhausts the set and thus has a probability of 1.

In our examples, we will mostly be talking about **mutually exclusive outcomes,** *meaning that if one outcome of an event occurs at a given time no other outcome can occur at that time.* For example, if an *H* occurs, a *T* cannot occur on the same toss. Our discussion will be further limited to **independent events,** *meaning that the outcome of one event will in no way affect the outcome of a subsequent event;* for example, getting an *H* on the first toss of a coin will in no way affect the outcome of a second toss.

We are interested in asking questions such as, ''What is the probability of getting exactly two *H*'s in three tosses?'' To answer this question, we simply compute the number of ways (arrangements of *H*'s and *T*'s) that two *H*'s can occur in three tosses, and form a fraction with this number over the total possible number of arrangements of *H*'s and *T*'s that can occur in three tosses. For example, if a coin is flipped three times, *H*'s and *T*'s can be combined into eight possible arrangements:

Possible Arrangements of *H*'s and *T*'s in Three Tosses		
Toss 1	Toss 2	Toss 3
H	*H*	*H*
H	*H*	*T*
H	*T*	*H*
H	*T*	*T*
T	*H*	*H*
T	*T*	*H*
T	*T*	*T*

Every specific arrangement of H's and T's is equally likely, so it can be represented by an equal area in the set or, equivalently, by an equal probability. Since there are eight possible arrangements, the probability of any one of them is 1/8. Thus, questions such as, ''What is the probability of obtaining exactly two H's in three tosses?'' can be answered by counting the number of arrangements containing the target (in this example, two H's) and multiplying by 1/8 (or, equivalently, dividing by 8). Since there are three equally likely ways (i.e., arrangements) of getting two H's (i.e., HHT, HTH, THH), the probability would be 3/8. You can also see that if you shade the areas represented by these three arrangements, you will shade 3/8 of the total area (see table). The relationship between probability and areas is important, and we will discuss it further in the next chapter.

We can use a formula to find the number of arrangements that can be obtained for a given target (e.g., two H's) when the order of the arrangement (e.g., HHT, HTH, or THH) is not a factor. This formula is the formula for combinations (C), where **combination** *means arrangement where order is not important:*

combination:

arrangement of things where order within the arrangement is not important

$$C_K^N = \frac{N!}{(N-K)!K!} \qquad\qquad 4.1$$

where N is the number of events from which arrangements can be made (e.g., number of tosses), and K is a type of designated target (e.g., all outcomes having two H's). The symbol ! represents *factorial.* Factorial means to take the value of a number, say K, and multiply together the sequence $(K)(K-1)(K-2) \cdots [K-(K-1)]$. For example, if we want to know how many combinations there are for getting two H's in three tosses, N is 3 and K is 2, so our answer is

$$C_2^3 = \frac{3!}{(3-2)!2!} = \frac{3 \cdot 2 \cdot 1}{1 \cdot 2 \cdot 1} = 3$$

There are three combinations which we have already shown are HHT, HTH, and THH.

The total number of arrangements from a sequence of events, each event having two possible outcomes, is 2^N, where N is the number of events in the sequence. For example, if the number of tosses is 3, then the total number of arrangements of H's and T's in three tosses is 2^3, or 8. **The probability of an outcome with a particular number of H's (or T's) is given by the ratio of the total number of equally likely arrangements containing the target number of H's (or T's) over the total number of possible arrangements.** (An equivalent way of computing this probability is to find the probability of one arrangement and multiply that by the number of equally likely arrangements.) The probability of getting exactly two H's in three tosses is then

$$\frac{C_2^3}{2^3} = \frac{3}{8} \quad \text{or} \quad \frac{1}{8} \times 3 = \frac{3}{8}$$

Let us do another example. What is the probability of getting exactly six H's in eight tosses?

First, find the number of ways that six H's can be arranged within eight tosses. This can be found with the combinations formula with $N = 8$ and $K = 6$:

$$C_6^8 = \frac{8!}{(8-6)!6!} = \frac{8\cdot7\cdot6\cdot5\cdot4\cdot3\cdot2\cdot1}{2\cdot1\cdot6\cdot5\cdot4\cdot3\cdot2\cdot1} = 28$$

Second, divide this number of arrangements by the total number of possible arrangements (i.e., 2^N). In our example, $N = 8$, so

$$2^N = 2^8 = 2\cdot2\cdot2\cdot2\cdot2\cdot2\cdot2\cdot2 = 256$$

The probability of getting exactly six H in eight tosses is therefore 28/256.

Additional Terminology

You must be sensitive to the terms used in probability:

1. *exactly:* this number only
2. *at least:* this number and all higher numbers
3. *less than:* all numbers below a specific number
4. *greater than:* all numbers above a specific number
5. *not:* all numbers except a specifically stated number
6. *between:* numbers between but not including two numbers

You can keep these terms straight if you think of a set of outcomes such as those in rolling a die:

number on face of die

1 2 3 4 5 6

Each outcome (i.e., number on face of die that is ''up'') is equally likely. Now, ''rolling a die and getting a number greater than 4'' means outcomes 5 and 6. ''At least 4'' means outcomes 4, 5, 6. ''Not 4'' means outcomes 1, 2, 3, 5, 6. ''Exactly 4'' means outcome 4. ''Less than 4'' means outcomes 1, 2, 3. ''Between 4 and 6'' means outcome 5.

A helpful relationship is

$$P(A) = 1 - P(\textbf{not } A)$$

This is read as, ''the probability of outcome A is one minus the probability of not getting outcome A.'' Therefore, the probability of rolling a die and getting a 4 is 1/6. The probability of not getting a 4 is $1 - 1/6 = 5/6$, which is the same as the probability of getting a 1 or 2 or 3 or

5 or 6, which is 1/6 + 1/6 + 1/6 + 1/6 + 1/6 = 5/6. You can frequently save work by using the formula.

When outcomes of events are independent and mutually exclusive, as in our examples thus far, a helpful rule is the **Additive Rule: When outcomes are linked by ''or,'' or can be linked by ''or,'' add their probabilities.** For example, what is the probability of getting three H or three T in three tosses? The probability of three H is 1/8. The probability of 3 T is 1/8. So the probability of three H or three T is 1/8 + 1/8 = 2/8, or 1/4. Note that there are equivalent ways of stating the same problem. For three tosses, the following questions are equivalent: What is the probability of all H or all T? What is the probability of getting more than two H or less than one H? What is the probability of all the coins showing the same face? What is the probability of not getting at least one H or one T? What is the probability of not getting one H or two H or one T or two T?

As we will soon see, of particular relevance to statistical tests are probability problems involving the notion of *at least.* For example, what is the probability of getting at least 8 H in 10 tosses? An equivalent way of saying this is, ''What is the probability of getting 8 H or 9 H or 10 H in 10 tosses?'' Thus, we find the probability for 8 H, for 9 H, and for 10 H, then add them together to answer the question. The answer is

$$\frac{45}{1024} + \frac{10}{1024} + \frac{1}{1024} = \frac{56}{1024} \quad \text{or} \quad .0055$$

BOX 4.2 (optional)
Non-Mutually-Exclusive Two-Target Situations

Suppose you are asked to find the probability of drawing an ace or a red card with one draw from a standard well-shuffled deck of 52 playing cards. The probability of a red card is 26/52, and the probability of an ace is 4/52. If you use the additive rule, you get 26/52 + 4/52 = 30/52, which is wrong. Why? Because the two targets were not mutually exclusive. You can get a red card that is also an ace. Thus, the red aces are being counted twice when you add the two separate probabilities. Thus you must subtract the probability of the joint occurrence of the outcomes of two events from their combined probability sum. Since there are two red aces, the answer to the question is 26/52 + 4/52 − 2/52 = 28/52. What we have described can be stated in algebraic form as

$$P(A) \text{ or } P(B) = P(A) + P(B) - P(A \text{ and } B)$$

This is read as, ''the probability of A or B equals the probability of A plus the probability of B minus the probability of the joint occurrence of A and B.''

When outcomes of events are mutually exclusive, $P(A$ and $B) = 0$, so we are left with $P(A)$ or $P(B) = P(A) + P(B)$. One can get the correct answers in other ways to the type of question being asked here. However, getting answers by using the given formula will be helpful as a way of thinking about probability that will generalize to other situations in more advanced statistics topics.

Let's work some more problems.

Problem 1

What is the probability of drawing either an even-numbered card (i.e., 2, 4, 6, 8, or 10) or a club from a shuffled deck with one draw?

$$P(\text{club}) = \frac{13}{52}$$

$$P(\text{even card}) = \frac{20}{52}$$

$$P(\text{even numbered club}) = \frac{5}{52}$$

Therefore, $P(\text{club or even numbered card}) = \frac{13}{52} + \frac{20}{52} - \frac{5}{52} = \frac{28}{52}$

Problem 2

What is the probability of getting a club or spade from a shuffled deck in one draw?

$$P(\text{club}) = \frac{13}{52}$$

$$P(\text{spade}) = \frac{13}{52}$$

$$P(\text{both club and spade}) = \frac{0}{52}$$

$$P(\text{club or spade}) = \frac{13}{52} + \frac{13}{52} - \frac{0}{52} = \frac{26}{52} \quad \text{or} \quad \frac{1}{2}$$

Event Sequences

The probability of a series of outcomes of events occurring in a *specified sequence* is given by *multiplying* the probabilities of the outcome of each event. Thus, if the event is coin tossing and the designated outcome is H on toss 1, T on toss 2, H on toss 3, the probability of the sequence HTH is $1/2 \times 1/2 \times 1/2 = 1/8$.

Is there something familiar here? Hopefully, yes. We earlier saw that each arrangement of H and T with three tosses was equally likely, and since there are eight possible arrangements the probability of any particular arrangement (i.e., sequence) must be 1/8. In fact, the probability of any sequence of outcomes of events, each event having two equally likely outcomes is $(1/2)^N$, where N is the number of events (e.g., tosses in the sequence). Thus $(1/2)^N = 1/2^N$, and 2^N was the denominator in our earlier coin-toss problems. Often, choosing a strategy for working probability problems is like choosing a mate: keep trying until you find one that you like and will work.

Problem 3

What is the probability of your getting a spade, then an ace, then a red card in a sequence of three draws when you replace each card and reshuffle the deck after each draw? Notice that this is a sequence problem and must be worked differently than the previous problem.

$$P(\text{spade, ace, red card}) = \frac{13}{52} \cdot \frac{4}{52} \cdot \frac{26}{52}$$

$$= \frac{1}{4} \cdot \frac{1}{13} \cdot \frac{1}{2} = \frac{1}{104}$$

An outline for solving the type of probability problems that are found in this chapter is the following:

a. Determine if the problem involves a *specified sequence* (i.e., arrangement of outcomes) of independent, mutually exclusive outcomes of events. If so, multiply the probability of the outcomes of the separate events together.

b. Does the problem call for a two-target designation (e.g., club or ace), constituting a specific outcome from *one event* (e.g., drawing a card). If so, use the formula where you add the separate outcomes and then subtract the joint outcomes: $P(A \text{ or } B) = P(A) + P(B) - P(A \text{ and } B)$. (*Note.* This formula for finding joint occurrences is only appropriate for two designated targets of an outcome (e.g., a club or an ace). The same principle holds for more than two designated targets in a more complicated fashion that is beyond the purpose of this section.)

c. Does the problem call for finding the outcome of one event where the outcome can be *any number of independent, mutually exclusive targets* (e.g., getting a 7, or 10, or an ace, etc, in one draw)? If so, the probability of the outcome is the sum of the probabilities of separate targets within the event: $P(A) \text{ or } P(B) \text{ or } P(C) \ldots = P(A) + P(B) + P(C) \ldots$. When $A, B, C \ldots$ represent specific arrangements of an outcome (e.g., an unknown number of arrangements of six H in 10 tosses), the combinations formula is appropriate for finding the number of arrangements:

$$C_K^N = \frac{N!}{(N - K)(K)!}$$

Don't despair at the seeming complexity. Working some exercises will enormously simplify things. Also, the soon to be used *sign test* will use *only* the principles involved in determining probabilities of sequences of coin tosses.

Example 2—continued. Let us go back to the situation where your friend has been winning more than her fair share of bets, 10 out of 11 to be exact. We can now compute step 2 in the decision-making process. Under honest conditions, the probability of losing 10 out of 11 would be

$$\frac{C_{10}^{11}}{2^{11}} = \frac{11!/(1!)(10!)}{2^{11}} = \frac{11}{2048}$$

or roughly 1 in 200.

If your criterion for accepting an outcome of an event as chance was 1 in 20 (step 1), then clearly your friend would be suspected of cheating (step 3). (As we will see in the next example, a more technically correct way of stating the problem is, "What is the probability of losing *at least* 10 out of 11 tosses?") Notice, however, that you did not prove in an absolute sense that your friend was cheating, only that it was quite likely. **This is the usual case for all statistical tests. Nothing is proven in an absolute sense (i.e., with a probability of 1), but the probability associated with the "by chance" explanation can be so small that one accepts an alternative as being the true situation.**

This situation is very similar to one involving a jury that has to make a decision based on circumstantial evidence. Suppose a suspect has been accused of murder, but no body has been found. Conceivably the victim could have run away or wandered away due to an amnesic episode or some other reason not involving foul play. However, suppose the alleged victim was last seen in the presence of the accused, the accused had a history of violence, and the victim was known to have decided to break off a relationship with the accused on the night of the victim's disappearance. Now before the foul-play hypothesis can be seriously pursued, the hypothesis that a "normal" event has happened has to be rejected even though it cannot be conclusively proven that a normal event has not happened. With the rejection of the normal-event hypothesis, one looks for an alternative explanation. Then, in the present example, if the jury thinks that the likelihood is high enough that the suspect was responsible for an assumed murder, the suspect will be convicted.

THE NULL HYPOTHESIS AND STATISTICAL INFERENCE

null hypothesis:

how characteristics of data would be expected to look in a population when the independent variable has no effect

What we have been calling the "by chance" explanation is associated with what is called the *null hypothesis* (H_0). **The term *null* means *no* in this context and refers to *how data or characteristics of data would be expected to look in a population when the independent variable has no effect*.** Since data from experiments are almost always samples, not populations, the possibility of sampling error because of random influences is always present in this sampled data. Thus, when the null hypothesis is true, any particular sample characteristic may be different from the corresponding population characteristic because of sampling error (i.e., random influences). A statistical test of inference will provide a probability value indicating the likelihood that a particular sample or characteristic of a sample has occurred under conditions where the null hypothesis was true. The experimenter then uses a decision criterion (usually 1 in 20) whereby he or she rejects H_0 if the probability is less than 1 in 20 that the characteristic of interest in the sample has occurred due to random influences. When H_0 is rejected, it can be inferred that something other than random influences is operating on the data. This ''something'' is usually inferred to be the independent variable.

THE SIGN TEST

Example 3. The principles that we have learned can be applied to an experimental situation. Suppose we are concerned about driving under the influence of drugs and we want to test to see if a particular drug will slow down reaction time. We therefore obtain a device that will measure in milliseconds the time that it takes to depress a key after a light is flashed. We find out that people vary up to about .10 seconds from trial to trial in their reaction times under a no-drug condition (if there was no variability, statistics would not be needed). Now, we set up a situation where we can measure the reaction times of 11 different people under both drug and no-drug (placebo) states. For simplicity we will assume that there are no practice effects and a sufficient time lapse is given so that there are no carryover effects from the drug. We then test each person in the drug and no-drug conditions and subtract their scores in the placebo condition from their scores in the drug condition to yield plus or minus values as shown in Table 4.1.

Now assume that the drug has absolutely no effect on reaction time. Effectively, this would mean that we were measuring a person's reaction time on two different occasions in a no-drug state. If we were to subtract a person's score in the placebo condition from his or her score in the ineffective drug condition, and looked at a population (i.e., an infinite number) of such scores, we would expect to see an equivalent number of pluses and minuses. Thus, H_0 in this situation would be an equal number of pluses and minuses. However, the number of pluses and minuses in our sample of 11 cases may not be exactly equivalent (in this situation a 6:5 split) because of random influences. As the number of pluses or minuses gets larger than the number designating equivalence, the probability gets smaller that the arrangement is due to random influences only. Notice that there is a very reasonable alternative for many pluses, namely, that the independent variable is effective.

The logic of the sign test is the following. If we assume *random variation from occasion to occasion,* we would assume some people to be faster during the first test and some people to be faster during the second test. Each person would have a 50-50 chance of being fastest on the first test (also on the second test). Therefore if we measure reaction time in hundredths of a second and subtract each person's first test (i.e., no-drug) score from their second test (i.e., ineffective drug) score, sometimes we would have positive values and sometimes negative values as shown in Table 4.1.

TABLE 4.1
Hypothetical Drug Study Data

Sample	No Drug (P)	Ineffective Drug (D)	$D - P$	
S_1	33	35	$+2$	
S_2	31	29	-2	
S_3	28	29	$+1$	(random
S_4	27	30	$+3$	arrangement
S_5	33	27	-6	of pluses
S_6	30	32	$+2$	and minuses)
S_7	31	34	$+3$	
S_8	30	28	-2	
S_9	29	27	-2	
S_{10}	31	28	-3	
S_{11}	30	31	$+1$	

If the drug were ineffective, we would expect a random arrangement of pluses and minuses across the 11 scores when each subject's score in one condition is subtracted from his or her score in the other condition. However, if the drug were slowing people's reaction times, we would expect to see larger numbers in the drug condition than in the placebo (no-drug) condition, resulting in many more plus numbers than minus numbers when the placebo condition scores are subtracted from the corresponding drug condition scores. We can therefore make a decision about the effectiveness of the drug in the same way that we make a decision about cheating in the coin-toss situation. Let us place this example within our three-step outline for decision making that has been modified to include H_0.

α level:

probability criterion for rejecting or retaining the null hypothesis

Step 1.   Determine the probability criterion level whereby H_0 for the outcome of a set of data will be rejected. In most experiments this criterion level is set by convention to be 1 chance in 20 (i.e., .05). *This probability criterion is also called the α level.* (α is pronounced *alpha.*)

Step 2.   Determine the probability that the event (i.e., the results of the experiment) happened by chance (i.e., random influences).

Step 3.   Decide whether to reject H_0.

We will now elaborate on step 2 and lay out the procedure for determining the probability of outcomes, assuming that difference scores (e.g., subtracting scores in one condition from scores in another) have a 50-50 chance of being "plus" scores or "minus" scores. The idea is to find a cutoff that separates outcomes that one will consider to be chance outcomes from ones that will be considered too rare to have occurred by chance. If our criterion for rejecting the H_0 is 1 in 20 (.05), then we need to find out how many pluses (or minuses) would have to occur within a set before we would reject H_0. We use the procedure in Box 4.3, using a directional test. (We will shortly explain the difference between a directional and a nondirectional test.)

Step 1.   Determine the probability of the most extreme possible outcome. Then compare this probability with the criterion probability (usually .05). If it is less than the criterion probability, go to step 2.

Step 2.   Find the probability of the second most extreme outcome and add it to the most extreme. If this combined probability is less than the criterion, then find the probability of the third most extreme and add it to the more extreme probabilities and compare to the criterion.

Step 3.   Continue the procedure until a probability is obtained that is larger than the criterion. The cutoff for rejecting H_0 will be between the outcome having a probability above the criterion and the next most extreme outcome having a probability less than the criterion.

Let's clarify this procedure by working out our drug-study example. First, the probability of the most extreme outcome (11 pluses) is

$$\frac{C_{11}^{11}}{2^{11}} = \frac{11!/(0!)(11!)}{2^{11}} = \frac{1}{2^{11}} = \frac{1}{2048}$$

(Note: 0! is defined to be 1.) The computed probability is 1/2048, which is less than .05.

Second, the probability of the next most extreme outcome (10 pluses) is

$$\frac{C_{10}^{11}}{2^{11}} = \frac{11!/(1!)(10!)}{2^{11}} = \frac{11}{2^{11}} = \frac{11}{2048}$$

And the more extreme probability (e.g., 1/2048):

$$\frac{11}{2048} + \frac{1}{2048} = \frac{12}{2048} < .05$$

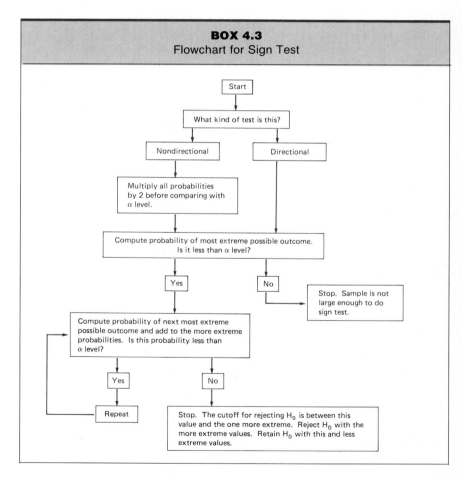

BOX 4.3
Flowchart for Sign Test

so continue with nine pluses:

$$\frac{C_9^{11}}{2^{11}} = \frac{55}{2048}$$

Add the more extreme probabilities (e.g., 12/2048):

$$\frac{55}{2048} + \frac{12}{2048} = \frac{67}{2048} < .05$$

so continue with eight pluses:

$$\frac{C_8^{11}}{2^{11}} = \frac{165}{2048}$$

Add the more extreme probabilities (e.g., 67/2048):

$$\frac{165}{2048} + \frac{67}{2048} = \frac{232}{2048} > .05$$

The cutoff is therefore between eight pluses and nine pluses. With eight or fewer pluses we retain H_0. With nine or more pluses we reject H_0. Therefore, for the data in Table 4.1, we retain H_0. If the actual data looked like Table 4.2, we have nine plus scores, so we reject H_0. Then, assuming we have an experimental design that is solid (i.e., no extraneous variables creating differences between conditions), we infer that the drug is accounting for the large number of pluses (i.e., higher scores in the drug group). We then conclude that the drug slows down reaction time.

TABLE 4.2
Hypothetical Drug Study Data

	No Drug (P)	Drug (D)	$D - P$
S_1	30	28	-2
S_2	29	33	$+4$
S_3	31	34	$+3$
S_4	23	25	$+2$
S_5	29	31	$+2$
S_6	33	32	-1
S_7	28	30	$+2$
S_8	31	35	$+4$
S_9	33	36	$+3$
S_{10}	29	33	$+4$
S_{11}	28	34	$+6$

DIRECTIONAL VERSUS NONDIRECTIONAL TESTS

directional test:

test wherein H_0 is rejected only if the outcome of a set of data is in a specified direction

nondirectional test:

test whereby a decision about whether to reject H_0 is made independent of the direction of the outcome of the data in the test

In the previous example, we were interested in a directional question. *By a **directional question**, we mean an outcome that is of interest only if a deviation from chance is in a specified direction.* With **nondirectional** *questions, we are interested in outcomes that deviate from chance irrespective of direction.* To illustrate, in the coin-toss example, we were only interested in whether or not someone was cheating against you, not whether someone was cheating for you by having you win more than you should by chance. Therefore, we used a directional test. Likewise, in the drug-study example we only considered the situation where the drug might slow down reaction time, not speed it up, and therefore we used a directional test. If we ask the questions, ''Did someone cheat either for you or against you?'' or ''Did the drug either slow down *or* speed up reaction?'' we are asking a *nondirectional* question. With a nondirectional sign test, the probability of a chance occurrence in step 2 is computed identically to that of a directional test *except that one must multiply the obtained probability by 2 before a comparison with the criterion probability is made.* To illustrate with the previous sign test data, the probabilities at the corresponding stages in finding the cutoff for rejecting H_0 would be

$$2 \left[\frac{C_{11}^{11}}{2^{11}} \right] = \frac{2}{2048} < .05 \quad \text{so continue}$$

$$2 \left[\frac{C_{11}^{10}}{2^{11}} \right] = \frac{22}{2048} + \frac{2}{2048} = \frac{24}{2048} < .05 \quad \text{so continue}$$

$$2 \left[\frac{C_{11}^{9}}{2^{11}} \right] = \frac{110}{2048} + \frac{24}{2048} = \frac{134}{2048} > .05 \quad \text{stop}$$

The cutoff is between 9 and 10 pluses or minuses instead of between 8 and 9 pluses as with the directional test.

The reason that the nondirectional test probabilities are twice that of the directional test probabilities is that the probability for each corresponding opposite extreme value is the same. Under random arrangements of pluses and minuses, the probability of 11 pluses is exactly the same as that for 11 minuses, that of 10 pluses is equal to 10 minuses, and so on. With a nondirectional test, one is in effect saying, ''What is the probability of this outcome (e.g., X or more pluses) or this outcome (e.g., X or more minuses) occurring?'' As we saw in the pobability section, an *or* statement means that we are adding probabilities. Adding two equal values is the same as multiplying one of them by 2.

A discussion of the choice concerning which test should be used under what conditions can be found in Chapter 12. The bottom line of that discussion is that, with nearly all empirical research, the non-directional test should be used.

DECISION THEORY

decision theory:

focuses on the types of errors that can be made in situations where a yes or no decision has to be made and the choice can be correct or incorrect

Decision theory involves uncertain situations where a yes or no decision has to be made and where either choice may be correct or incorrect. One common situation where decision theory is appropriate is in jury trials. A jury may be correct in two ways. A jury may find a person innocent when, in fact, he or she is innocent. Or, a jury may find a person guilty when, in fact, he or she is guilty. There are also two ways that juries can make mistakes. They may say a person is innocent when he or she is guilty, or they may say a person is guilty when he or she is innocent. We can illustrate this in the following matrix:

		Actual Truth	
		Innocent	Guilty
Decision	Innocent	Correct	Error
	Guilty	Error	Correct

In all situations like this, the decision criterion creates an inverse relation between the types of errors such that as the probability of one type gets smaller, the probability of the other type gets larger. Which type of error gets large or small relative to the other depends on the criteria of the jury for guilt or innocence. For example, if the jury wants a small probability of making an error whereby an innocent person is convicted (i.e., they don't want to send an innocent person to jail), then an overwhelming amount of evidence against the accused will have to be presented for a conviction to be obtained. However, with this very stringent criterion for convictions, the other kind of error increases and more guilty people will be set free. Conversely, if a lower criterion is established (e.g., ''Let's get the criminals off the street''), then more innocent people will be put in jail. Our society tries to minimize errors whereby innocent people are convicted (a person is innocent until proven guilty beyond reasonable doubt), whereas other societies, particularly with political suspects, try to minimize errors whereby guilty people are set free.

Let us take another example: Suppose that you wish to decide whether to marry someone. The matrix would look as follows:

Truth About Potential Mate

		Winner	Loser
Marriage Decision	Marry	Correct	Error
	Stay Single	Error	Correct

Which type of error do you think is more serious and how would that affect your decision criterion? Many practical situations can be placed within the decision theory framework. See Box 4.4 for a comparison of

---

BOX 4.4
Comparison of Examples 2 (Coin-Toss) and 3 (Drug-Study)
Using Decision Theory

Decision*/Actuality	Outcome
Retain H_0/H_0 is true Ex. 2: H_0 = random number of wins and losses Ex. 3: H_0 = drug is ineffective; thus random pluses and minuses in sample	Ex. 2: You correctly conclude that your friend was not cheating you. Ex. 3: You correctly conclude that the drug is ineffective.
Retain H_0/H_0 is false	Ex. 2: Error; you decide that your friend is not cheating you when in fact she was. Ex. 3: Type II error; you decide that the drug was not effective, when in fact it was.
Reject H_0/H_0 is false	Ex. 2: You correctly conclude that your friend was cheating you. Ex. 3: You correctly conclude that the drug would slow down reaction time.
Reject H_0/H_0 is true	Ex. 2: Error; you accuse your friend of cheating when in fact she was not. Ex. 3: α or Type I error; you claim that the drug was effective when in fact it was not.

* Criterion for decision: Ex. 2—entirely subjective, the relative severity of the two types of errors should be made before choosing criterion; Ex. 3—set by convention to be .05 for Type I or α errors.

Examples 2 and 3 in this chapter using decision theory. A major plus for this approach is that it forces one to more fully examine alternative consequences. With respect to retaining or rejecting H_0 in experiments, the same situation applies and can be illustrated:

Actuality

	H_0 Is True	H_0 Is False
Retain H_0	Correct	Error (Type II or β)
Reject H_0	Error (Type I or α)	Correct

Decision

α **error :**

(i.e., Type I error) error made by rejecting a null hypothesis that is true

β **error :**

(i.e., Type II error) error made by retaining a null hypothesis that is false

In situations involving experiments, the two types of decision-making errors have names. *An α error (i.e., Type I error) is the error that one would make if one rejected H_0 when, in fact, H_0 was true. A β error (i.e., Type II error) is the error that one would make if one retained H_0 when, in fact, H_0 was false. The α level set by a researcher determines the probability of an α error.* If the α level were set at .20, then there would be a 20% chance that the researcher would reject H_0 when it was true. In research, the most serious error is considered to be the Type I error. Therefore, the probability of committing such an error is kept small by setting it usually at 1 in 20 or .05. In other words, an experimenter wants to be reasonably sure that the independent variable is indeed effective (assuming a well-controlled experiment) before he or she says it is.

SUMMARY

The purpose of statistical tests of inference is to determine the probability that differences that exist between sets of data are entirely due to random or chance influences. If this probability value is less than a chosen criterion, then the investigator will reject chance as the reason for the differences that exist between sets of data. This "by chance" explanation is associated with the *null hypothesis* (H_0). The term *null* means *no* in this context and refers to how data or characteristics of data would be expected to look in a population when the independent variable has no effect. Since data from experiments are almost always samples, not populations, the possibility of sampling error due to random influences is always present in this sampled data. Thus, when the null nypothesis is true, any particular sample characteristic may be different from the corresponding population characteristic because of sampling error (i.e., random influences). A statistical test of inference will provide a probability value indicating the likelihood that a particular sample or characteristic of a sample has occurred under conditions where the null hypothesis was true. The experimenter then uses a decision criterion whereby he or she rejects H_0 if the probability is less than 1 in 20 that the characteristic of interest in the sample has occurred due to random influences. When H_0 is rejected, it can be inferred that something other than random influences is operating on the data. This ''something'' is usually inferred to be the independent variable.

Three steps are involved in the process of deciding whether to reject H_0.

Step 1. Establish a probability criterion (also called α level) for rejecting H_0. In most experiments, the α level is set at .05.

Step 2. Assuming H_0 to be true, determine the probability that the outcome of a particular event, such as scores resulting from an experimental manipulation, has happened entirely because of random influences.

Step 3. Make a decision about rejecting H_0 based on a comparison of the decision criterion in step 1 with the probability value obtained in step 2. If the probability value in step 2 is *smaller* than the criterion established in step 1, then *reject H_0*. If the probability value in step 2 is *larger* than the criterion established in step 1, then *retain H_0*.

Probability is defined as the likelihood of an event occurring, and is measured on a scale ranging from 0 (it could not happen) to 1 (it definitely will happen). Probability values can be determined based on the relative frequency with which an event has occurred in the past or by *a priori* methods, whereby probabilities are determined based on facts surrounding the event. Our probability discussion involved discrete outcomes of events such as coin tosses. The total range of outcomes is called a *set* and is represented by a probability of 1. An outcome of event is mutually exclusive if only one outcome can occur at a time. Outcomes of events are independent if the outcome of one event in no way affects the outcome of a subsequent event.

Probabilities can often be found by taking a ratio of the number of ways a target outcome can occur over the total number of outcomes for the event, assuming that all outcomes are equally likely. The formula for *combinations* (C) gives the number of ways that things can be arranged when order isn't important:

$$C_K^N = \frac{N!}{(N-K)!K!}$$

When events are *independent* and *mutually exclusive,* linking outcomes by *or* means to add probabilities. A helpful relationship is

$$P(A) = 1 - P(\text{not } A)$$

When events are *not* mutually exclusive, then

$$P(A) \text{ or } (B) = P(A) + P(B) - P(A \text{ and } B)$$

The probability of a series of independent outcomes occurring in a specified sequence is given by multiplying the probabilities of the separate outcome of each event.

In the *sign test* the null hypothesis states that persons' scores in one condition when subtracted from their scores in another condition (i.e., difference scores) will result in an equivalent number of pluses and minuses in a population (i.e., infinite number of scores) and a random arrangement of pluses and minuses in a sample. The probability of any arrangement of pluses and minuses can be computed from the formula

$$\frac{C_K^N}{2^N}$$

as the basis of the computation. The computed probability value is determined in step 2 of the three-step process in determining whether to reject H_0.

Statistical tests can be directional or nondirectional. A *directional* test is appropriate for testing whether the deviation from chance is in a specified direction; for example, scores in group A can be larger than group B or smaller than group B, not both. A *nondirectional* test is appropriate when there is interest in one group of scores being either larger or smaller than another group.

Decision theory involves uncertain situations where a yes or no decision has to be made and either choice may be correct or incorrect. For example, an investigator may correctly retain H_0 or correctly reject H_0. The investigator might also reject H_0 when it is true (i.e., a *Type I* (α) *error*) or retain H_0 when it is false (i.e., a *Type II* (β) *error*). The two types of errors, α and β, are inversely related to one another; in other words, adopting a decision criterion that reduces the probability of one type of error will increase the probability of the other type of error. The α level set by a researcher determines the probability of an α error. Thus, in most empirical research, the probability of an α error under conditions where H_0 is true is .05.

KEY TERMS

event	.05 α level
outcome	decision theory

probability

mutually exclusive

independent

arrangement

combination

sequences of outcomes

null hypothesis

Type I or α error

Type II or β error

sign test

directional tests

nondirectional tests

Verbalize the procedure for:

(a) using the combinations formula

(b) finding the probability of exactly K heads in N tosses

(c) finding the cutoff point for the sign test

(d) retaining or rejecting H_0

PROBLEMS

1. With one role of an honest die, what is the probability (p) of getting

    a   a 1 or 6

    b   an even number

    c   not a number above 4

    d   a number less than 6 and odd

    e   an odd or even number

    f   a number greater than 6

2. How many arrangements of seven H in nine tosses are there? How many total arrangements of H and T are there in nine tosses? What is the p of getting exactly seven H in nine tosses?

3. What is the p of getting at least seven H in nine tosses?

4. What is the p of getting seven or more H or T in eight tosses?

5. What is the p of getting less than six H in seven tosses?

6. Refer to Box 4.2. From a shuffled deck of regular playing cards with replacement, what is the p of

    a   an ace or deuce in one draw

    b   five red cards in a row in five draws

    c   a red card, then a black card in two draws

    d   getting exactly three red cards in four draws (can be worked as in coin tosses)

    e   getting at least six black cards in seven draws

    f   getting all cards of the same color in three draws

    g   getting three clubs in three draws

    h   not getting a black or red card in one draw

    i   getting a club or ace in one draw

    j   getting a king or red card in one draw

    k   getting a face card (jack, queen, king, ace) or spade in one draw

7a. A coin is flipped eight times:

    a   What is the p of four H or four T?

    b   What is the p of less than seven H?

    c   What is the p of more than seven H or less than one H?

    d   How many total arrangements of H and T are there?

    e   Is an odd number of H's and an even number of H's equally likely? Why or why not?

7b. Answer the five questions in 7a using nine coin flips.

8. Identify and then discuss the relative severity of the two types of decision-making errors within each of the following situations. You need not specify whether the errors are Type I or Type II.

    a   Should a new drug with unknown but perhaps harmful side effects be allowed to be sold if there is evidence that it can cure the flu?

    b   Should you wear seat belts on a short trip?

    c   Should you invest your life savings in a small but promising computer company?

9. What would be the cutoff for (a) a directional sign test with $N = 8$, $\alpha = .05$; (b) a nondirectional sign test with $N = 8$, $\alpha = .05$?

10. What would be the cutoff for (a) a nondirectional sign test and (b) a directional sign test with $N = 10$, $\alpha = .05$?

11. What is the smallest number of subjects that one could use and have the possibility of rejecting H_0 with a nondirectional sign test using $\alpha = .05$?

12. Is it easier to reject H_0 with a directional test than with a nondirectional test? Comment.

13. For $\alpha = .05$, should one reject H_0 using a nondirectional test for the following data. Use the sign test and show your work.

P	D
33	31
29	30
36	33
32	35
35	31
36	32
38	34
40	31
31	30
28	27
34	32

14. If D was a drug condition and P was a placebo condition and the scores are reaction times in hundredths of a second, how would you interpret the results of the sign test in problem 13?

CHAPTER 5

Transformed Scores and the Normal Curve

TRANSFORMED SCORES

operational definition:

states the way in which an independent or dependent variable is measured

Suppose you are a judge at a state fair and are solely responsible for awarding the grand prize to the person who has grown the largest vine-grown vegetable of its type. All vine-grown types of vegetables are eligible for the competition. So now you have to decide whether Sally's 3-pound tomato is bigger for a tomato than Larry's 10-pound cantaloupe is for a cantaloupe. Is there a way to make an objective decision in this situation? Yes. First we need to give an operational definition to the term *largest*. We might try to use a measure like circumference, but that might prove difficult for vegetables that are more oblong than round. So we decide to use weight as the measure from which we will make a decision. Now, to decide between Sally and

Larry, we weigh all of the tomatoes and cantaloupes in the competition and find the mean weight for the tomatoes and for the cantaloupes. Suppose we find that the mean weight of the tomatoes was 2 pounds, whereas the mean weight of the cantaloupes was 8 pounds. Would it now be fair to give the prize to Larry because his cantaloupe was 2 pounds above the mean for cantaloupes, whereas Sally's tomato was only 1 pound above the mean for tomatoes? No, because we still don't know if 2 pounds above the mean is heavier for the cantaloupe than 1 pound above the mean is for the tomato. But suppose we now find the standard deviation for the weights of all of the tomatoes and all of the cantaloupes that had been entered. We have now both the mean and a measure of score spread for each vegetable. Thus, we will have a meaningful comparison between the cantaloupe and the tomato if we compute the number of standard deviations each is from the mean of its group. If the standard deviation for the tomatoes was 1/4 pound and the standard deviation was 1 pound for the cantaloupes, then Sally would win because her tomato was 4 standard deviations above the mean for tomatoes, whereas Larry's cantaloupe was only 2 standard deviations above the mean for cantaloupes.

The number of standard deviations from the mean can be the "common denominator" from which to compare things that are on very different scales. For example, one could use this technique to determine who was the better *per average* hitter, relative to peers of his time, Rod Carew, Wade Boggs, Ted Williams, or Ty Cobb. Although you may never have used this technique, you quite likely have had it used on you. For example, it is common in many colleges and universities to grade large classes "on the curve." When this is done, the scores from exams are *transformed* (i.e., converted) into *standard scores* by the procedures that we have described, and grades are determined by averaging together the standard scores for the exams and assigning grades on a preset basis such as 10% A's, 20% B's, and so on. Thus, your grade is a reflection of your performance relative to others who took the course at the same time. A **standard score** (also called a Z score) *where the data represent a population is thus defined:*

transformed score:

score that has been transformed from a value on one scale to a value on a different scale through a mathematical operation

standard score:

(i.e., a Z score) a score that has been transformed by the formula $Z = (X - \mu)/\sigma$ or $Z = (X - \bar{X})/S$

$$Z = \frac{X - \mu}{\sigma}$$

5.1

where Z = standard score
X = original score
μ = mean of original scores
σ = standard deviation of original scores

If the data from which the Z score is computed represent a sample instead of a population, S is substituted for σ and the formula becomes

$$Z = \frac{X - \bar{X}}{S}$$

5.2

Thus, the Z score for Sally's tomato was

$$Z = \frac{4 - 3}{.25} = 4$$

and the Z score for Larry's cantaloupe was

$$Z = \frac{10 - 8}{1} = 2$$

What we really did in this example was to compare the tomato and the cantaloupe by using Z scores.

RULES FOR TRANSFORMING SCORES

Let us more formally state the procedures that we used in computing a Z score, as well as other rules used in transforming scores. In all of the following cases, we start with a set of scores that we designate by X.

Rule 1: If a constant (any one number) k is added to each member of a set of scores, then

$$\bar{X}_{new} = \bar{X}_{old} + k$$

Example: Let $k = 5$ and $N = 4$.

X	$X + k$
4	9
1	6
2	7
5	10
$\Sigma X = 12$	$\Sigma X = 32$
$\bar{X}_{old} = 3$	$\bar{X}_{new} = 8$

Rule 2: If a constant k is subtracted from each member of a set of scores, then

$$\bar{X}_{new} = \bar{X}_{old} - k$$

Example: Let $k = 4$ and $N = 3$.

X	$X - k$
4	0
9	5
2	-2
$\Sigma X = 15$	$\Sigma X = 3$
$\bar{X}_{old} = 5$	$\bar{X}_{new} = 1$

Rule 3: If every score is multipled by a constant k, then

$$\bar{X}_{\text{new}} = k\bar{X}_{\text{old}}$$

Example: Let $k = 5$ and $N = 4$.

X	kX
5	25
2	10
1	5
4	20
$\Sigma X = 12$	$\Sigma X = 60$
$\bar{X}_{\text{old}} = 3$	$\bar{X}_{\text{new}} = 15$

Rule 4: If every score is divided by a constant k, then

$$\bar{X}_{\text{new}} = \bar{X}_{\text{old}}/k$$

Example: Let $k = 5$ and $N = 5$.

X	X/k
5	1
10	2
0	0
15	3
20	4
$\Sigma X = 50$	$\Sigma X = 10$
$\bar{X}_{\text{old}} = 10$	$\bar{X}_{\text{new}} = 2$

Rule 5: If a constant k is *added* to or *subtracted* from each member of a set of scores, then

$$\sigma_{\text{new}} = \sigma_{\text{old}}$$

Rationale: Adding or subtracting a single number to or from a set of scores will not change the standard deviation, since the score spread reflected by $X - \bar{X}$ will not change.

Rule 6: If every score within a set of scores is multiplied by a constant k, then

$$\sigma_{\text{new}} = k\sigma_{\text{old}}$$

Example: Let $k = 3$ and $N = 3$.

X	$X - \bar{X}$	$(X - \bar{X})^2$	kX	$X - \bar{X}$	$(X - \bar{X})^2$
2	-1	1	6	-3	9
1	-2	4	3	-6	36
6	$+3$	9	18	9	81

$\Sigma X = 9$ $\Sigma(X - \bar{X})^2 = 14$ $\Sigma X = 27$ $\Sigma(X - \bar{X})^2 = 126$

$\bar{X} = 3$ $\sigma_{old}^2 = 14/3 = 4.67$ $\bar{X} = 9$ $\sigma_{new}^2\ 126/3 = 42$

$\sigma_{old} = 2.16$ $\sigma_{new} = 6.48$

Rule 7: If every score within a set of scores is divided by a constant k, then

$$\sigma_{new} = \sigma_{old}/k$$

Rationale: Use previous example where $k = 3$ and $N = 3$.

X	X/k
6	2
3	1
18	6

$\sigma_{old} = 6.48$ $\sigma_{new} = 2.16$

STANDARD SCORES (Z SCORES)

The preceding rules allow us to transform sets of scores having different means and standard deviations into a scale with the same mean and standard deviation. The formula for standard scores (Z scores) given in Equation 5.1 is

$$Z = \frac{X - \mu}{\sigma}$$

When one converts all of the scores within a set into Z scores, the mean of the set of Z scores will always be 0 and the standard deviation will always be 1. *The mean of the Z scores will always be 0* because, from Rule 2, if a constant is subtracted from a set of scores then

$$\bar{X}_{new} = \bar{X}_{old} - k$$

With Z scores, $k = \bar{X}_{old}$, because in transforming a set of scores into Z scores the mean of that set (which is a constant) is subtracted from every score in the set.

So

$$\bar{X}_{new} = \bar{X}_{old} - \bar{X}_{old} = 0$$

The σ of a set of Z scores will always be 1, since, by Rule 7, if a set of scores is divided by a constant k,

$$\sigma_{\text{new}} = \sigma_{\text{old}}/k$$

Since $k = \sigma_{\text{old}}$,

$$\sigma_{\text{new}} \left(\sigma \text{ of } Z \text{ scores}\right) = \frac{\sigma_{\text{old}}}{\sigma_{\text{old}}} = 1$$

An example will illustrate these principles. We will start with a set of X scores, transform them into Z scores, and find the mean and σ of the Z scores.

X	$X - \bar{X}$	$(X - \bar{X})^2$	Z	$Z - \bar{Z}$	$(Z - \bar{Z})^2$
12	2	4	$(12 - 10)/2 = +1$	$+1$	$+1$
8	-2	4	$(8 - 10)/2 = -1$	-1	$+1$
8	-2	4	$(8 - 10)/2 = -1$	-1	$+1$
12	2	4	$(12 - 10)/2 = +1$	$+1$	$+1$
$N = 4 \quad \Sigma X = 40$		$\Sigma(X - \bar{X})^2 = 16$	$\Sigma Z = 0$		$\Sigma(Z - \bar{Z})^2 = 4$
$\bar{X} = 10$		$\sigma_x = \sqrt{16/4} = 2$	$\bar{Z} = 0$		$\sigma_z = \sqrt{4/4} = 1$

Again, the mean of the Z scores is 0, and the standard deviation is 1.

Example. Suppose you have taken exams in math, history, and psychology. Your scores on the three exams were 50, 65, and 90, respectively. On which exam did you do best *relative to the other students in the class?* We can answer this question by comparing your Z scores for each exam, provided we have the mean and standard deviation for each exam. We will list the necessary information and compute a Z score for each exam as in Table 5.1.

TABLE 5.1

	Math	History	Psychology
Score (X)	50	65	90
μ	30	90	70
σ	5	5	10
Z	$(50 - 30)/5 = 4$	$(65 - 90)/5 = -5$	$(90 - 70)/10 = 2$

Given the stated outcomes, you did best on the math exam, super terrific in fact. You also did great on the psychology exam, but bombed the history exam. Notice the importance of the sign in a standard score. A positive score means a score above the mean, while a negative score means a score below the mean. All of these Z scores are extreme. As you will soon see when you study the normal distribution, Z scores rarely go above $+2$ or below -2.

Z Scores and Shapes of Frequency Distributions

Converting a distribution of scores into Z scores will not affect the relative shape of the distribution. The new distribution may be narrower or wider, depending on how the new distribution is scaled, but all properties involving the shape of the distribution will be maintained. The Z score distribution will be an *exact* replica of the original distribution with the appropriate x axis scale. To demonstrate this, we will transform the scores in Table 5.2 into Z scores and draw line graphs to illustrate each set of scores.

TABLE 5.2

X	Z (fraction)	Z (decimal)
11	$-3/1.76$	-1.70
12	$-2/1.76$	-1.14
12	$-2/1.76$	-1.14
13	$-2/1.76$	-1.14
13	$-1/1.76$	$-.57$
13	$-1/1.76$	$-.57$
13	$-1/1.76$	$-.57$
13	$-1/1.76$	$-.57$
13	$-1/1.76$	$-.57$
14	0	0
14	0	0
15	$1/1.76$	$.57$
15	$1/1.76$	$.57$
15	$1/1.76$	$.57$
15	$1/1.76$	$.57$
16	$2/1.76$	1.14
17	$3/1.76$	1.70
17	$3/1.76$	1.70
17	$3/1.76$	1.70

As you can see the shapes of the frequency distributions in Figure 5.1 are identical. This is because converting raw scores into Z

Figure 5.1
A comparison of raw score and Z score distributions.

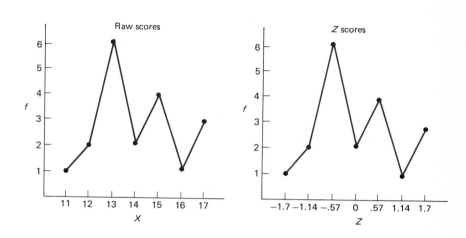

scores only changed the scale on the X axis. Nothing else changed. An analogy may help you understand this. If you are 6 feet tall and your little brother is 3 feet tall, you are twice as tall as he whether the measurement is in yards (2 versus 1) or feet (6 versus 3) or inches (72 versus 36). This kind of scale change does not affect proportionality.

TRANSFORMING STANDARD SCORES

T score:

score that has been transformed by the formula $T = 10Z + 50$

T Scores

There is one practical problem with the standard score. Most people, especially those not having a statistics background, are not used to thinking about the mean of a set of scores being zero, or a score of $+1$ being a very good score. The solution is to do a second transformation. After a set of scores are transformed into Z scores, one can then transform the Z scores onto a new scale. *Z scores that have been transformed into a scale having a mean of 50 and a standard deviation of 10 are called **T scores.*** A T score can thus be obtained from the formula

$$T = 10Z + 50 \qquad\qquad 5.3$$

To convert our previous exam scores for math, history, and psychology into T scores, we convert the raw (original) score or scores into standard (Z) score or scores, and then we convert each Z score into a T score by applying Equation 5.3. Since our Z score for our math exam was 4, the corresponding T score would be

Math	$T = 10(4) + 50 = 90$
for History	$T = 10(-5) + 50 = 0$
for Psychology	$T = 10(2) + 50 = 70$

The information contained in *T* scores is exactly the same as that contained in *Z* scores. The numbers are simply spread over a different, but equivalent, scale. A *T* score of 40 is interpreted in exactly the same way as a *Z* score of -1. A *T* score of 55 is equivalent to a *Z* score of $+.5$. Since about two thirds of the scores in a set (assuming the scores are normally distributed) fall between *Z* values of -1 and $+1$, about two thirds of the scores in a set fall between 40 and 60 when transformed into *T* scores.

SAT Scores

The transformation that you are probably most familiar with is the transformation of *Z* scores into Scholastic Aptitude Test (SAT) scores. *Z* scores are transformed into SAT scores by the formula

$$SAT = 100Z + 500$$

Thus, the mean of SAT scores is 500 and the standard deviation is 100. A SAT score of 600 is equivalent to a *T* score of 60. In fact, SAT scores are always 10 times their corresponding *T* score.

TRANSFORMED SCORES AND INFERENTIAL STATISTICS

The importance of transformed scores goes beyond the uses thus far presented. The idea behind the type of statistical test of inference that we will later study is to transform a value such as a difference between means (each mean representing an experimental condition) into a score that can be interpreted in a fashion similar to a *Z* score. Once we know how scores are expected to range throughout a distribution, we can ask probability questions such as, ''What is the probability that the tenth person you pass on campus between classes (a random choice) had a verbal SAT score of above 600?'' Within the context of an experiment, the question becomes, ''What is the probability that the transformed score that I have obtained from my statistical test has occurred under conditions where the null hypothesis was true?'' We then use that probability value in a fashion identical to the way that we used probability values in the sign test from Chapter 4.

THE NORMAL DISTRIBUTIONS

empirical distribution:

frequency distribution made up from an actual set of numbers or scores

The frequency distributions that one can compile from data of some type and illustrate with a histogram or frequency polygon are called *empirical frequency distributions. Empirical means based directly on some type of real data, be it height, SAT scores, reaction time scores, or whatever.* By contrast, a ***theoretical distribution*** *is not compiled from some data set, but its shape is determined by a mathematical function (i.e., equation).* Normal distributions are such distributions. They are a family (i.e., group) of distributions defined by a complex mathematical equation. Three of

theoretical distribution:

distribution defined by a mathematical formula (e.g., normal distribution or *t* distribution)

standard normal distribution:

a distribution that results when the scores from a normal distribution are transformed into *Z* scores

their shapes are displayed in Figure 5.2. If the scores within any normal distribution are transformed into *Z* scores with the usual *Z* transformation

$$Z = \frac{X - \mu}{\sigma}$$

then the resulting transformed distribution is called the *standard normal distribution*. As previously mentioned, when any set of scores is transformed into *Z* scores, the *Z* scores will have a mean of 0 and a standard deviation of 1. Therefore, the standard normal distribution also has a mean (μ) of 0 and a standard deviation (σ) of 1.

Figure 5.2
Normal distributions having three different standard deviations.

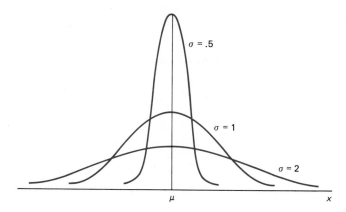

Of the millions of potential theoretical distributions, why do we study this particular family called the normal distributions? The reason is that the normal distributions are a very close match to many real-world empirical frequency distributions. For example, if you randomly select 100 males or females, measure their heights, weights, lengths of their big toes, and obtain scores on several measures such as running speed, weight lifting, and so on, you will find that the frequency distribution based on each of these measures has a shape very similar to a normal distribution.

There are three important characteristics of any normal distribution:

(a)   The normal distribution is unimodal.
(b)   The normal distribution is symmetric.
(c)   As the tails of the curve get farther away from the center, they get closer and closer to the *x* axis but *never* touch it.

AREAS UNDER THE STANDARD NORMAL DISTRIBUTION

Before we can put the normal distributions to work for us, we need to delineate the areas under the standard normal distribution. Once we have done this, we can answer questions involving areas of other normal distributions by transforming them into the standard normal distribution with the standard score formula (5.1) as previously given:

$$Z = \frac{X - \mu}{\sigma}$$

After discussing areas and relating them to probability, we will be in a position to explain how the normal distribution can be useful, as opposed to simply being a curiosity item of mathematicians.

We will start our discussion of areas and the normal distributions by assigning a value of 1 to the total area under the standard normal distribution. Just as a probability value can range from 0 to 1, so can an area under the standard normal distribution range from zero area to all (i.e., 100% or 1) of it. We now take advantage of the symmetry properties of the standard normal distribution and start marking areas from the center to one of the tails. This allows us to bypass the problem of where to start marking areas in a tail that is infinitely long. We next need a scale. Since numbers in the standard normal distribution represent scores transformed into Z scores, the mean (also midpoint) of the standard normal distribution is 0 and the standard deviation is 1. From this point, μ, we can scale in standard deviation units as shown in Figure 5.3. **These standard deviation units are called z units.** (We will follow the convention of using a lowercase z to refer to

Figure 5.3
Areas under the standard normal distribution.

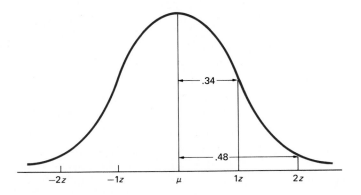

standard score units as applied to the standard normal distribution. An uppercase Z represents other standard scores.)

We now need to determine the area from the midpoint of the curve to various z's away from the midpoint. Enter our friends, the mathematicians. Using a technique called integral calculus, mathematicians have computed these areas for us, and the areas are displayed in Appendix B. Table B in Appendix B shows areas that have been marked off on the x axis in z units of hundredths. *The areas that are given in the table represent the area from the midpoint to so many z's from the midpoint.* Say we want to know the area from the midpoint to $1.42z$. We go down the z column until we come to 1.42. The value that we are looking for is .4222, as reproduced in Table 5.3.

TABLE 5.3
Proportion of Area under the Normal Curve between the Mean and z

z	
0.0	
.	
.	
.	
1.41	.4207
1.42	**.4222**
1.43	.4236

The value .4222 represents 42.22%, which is the percentage of the total area of the normal curve between the mean (μ) and 1.42 standard deviations (z's) from the mean. The shaded area in Figure 5.4 represents this z value area.

Notice that because of the *symmetry property of the normal curve,* we also know the following after finding the area between μ and $1.42z$ to be 42.22% of the total area:

(a) The curve between μ and $-1.42z$ is also 42.22% of the total area. (This symmetry is why only half of the areas under the normal curve are listed.)

(b) The area beyond $1.42z$ is 50% (the area for half of the curve) minus 42.22% = 7.78%, as shown in the unshaded part of the right half of Figure 5.4.

(c) The area below $1.42z$ (includes all negative z areas) is 50% (for the left half of the curve) plus the area between the mean and

1.42z, which is 42.22%. Thus, the area below 1.42z is 50% + 42.22% = 92.22%.

Figure 5.4
Area under the normal curve.

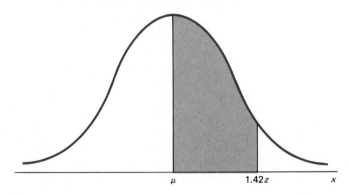

μ 1.42z x

Procedures for Finding Areas

All problems involving finding areas under the normal curve key on the following three words: *greater, less,* and *between* as illustrated.

> *Example 1.* What is the area beyond .39z? *Beyond* is equivalent to *greater than,* so the question can be rephrased as, what is the area for z values greater than .39? We start by shading the target area, and then we find from the z table that the area between μ and .39z is 15.17%. See Figure 5.5. The shaded area then equals 50% − 15.17% = 34.83%.

Figure 5.5
Area under the normal curve in Example 1.

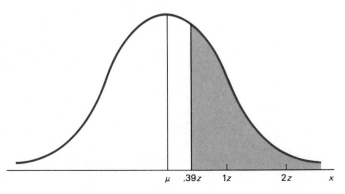

μ .39z 1z 2z x

Example 2. What is the area for a *z* value greater than − .39? Notice that this is *not* the same question as in Example 1, as Figure 5.6 shows. The shaded area is the target area, which is found by adding 15.17% to 50% to give 65.17%. The best way to prevent confusion about the area that you are looking for is to *sketch it.*

Figure 5.6
Area under the normal curve in Example 2.

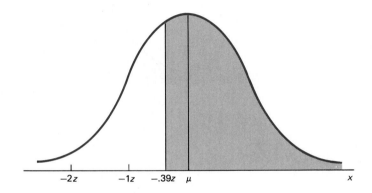

Example 3. What is the area below a *z* value of .70? The area between μ and .70z is 25.80%. Therefore, the target area is 50% + 25.80% = 75.80%. See Figure 5.7.

Figure 5.7
Area under the normal curve in Example 3.

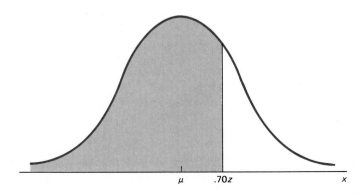

Example 4. What is the area below $-.70z$? By symmetry, the area between μ and $.70z = 25.80\%$. The target area is then $50\% - 25.80\% = 24.20\%$. See Figure 5.8.

Figure 5.8
Area under the normal curve in Example 4.

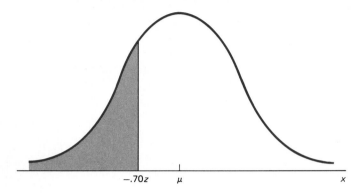

Example 5. What is the area *between* $1.80z$ and $1.21z$? Since the table only lists values starting from μ, we must find the area between μ and $1.80z$ and then subtract the area between μ and $1.21z$. The area between μ and $1.80z$ is 46.41%, and the area between μ and $1.21z$ is 38.69%. Therefore, our target area is $46.41\% - 38.69\% = 7.72\%$. See Figure 5.9.

Figure 5.9
Area under the normal curve in Example 5.

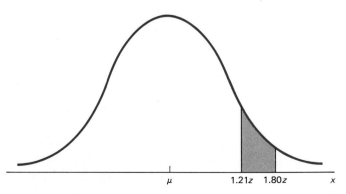

Example 6. What is the area *between* $-.31z$ and $.94z$? The area between μ and $.94z$ is 32.64% while the area between

μ and $-.31z$ is 12.17%. The target area is then 32.64% + 12.17% = 44.81%. Notice that if you tried to find the answer by first adding .94z to .31z, you would be very wrong because the area .31z's beyond .94 (the area between .94z and 1.25z) is much smaller than the area between μ and .31z. See Figure 5.10.

Figure 5.10
Area under the normal curve in Example 6.

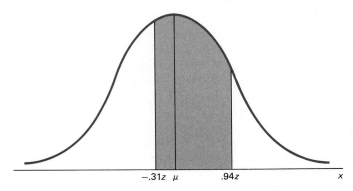

Example 7. How many z's do you have to go from μ to leave 5% of the area in a tail? This is a "backward" problem. You must look inside the normal curve table until you find a value that encompasses 45% (i.e., 50% − 5% = 45%) of the area between μ and the unknown z value. If the exact value isn't found, we use the closest value, which is 45.05%. The answer is then z = 1.65. See Figure 5.11.

Figure 5.11
Area under the normal curve in Example 7.

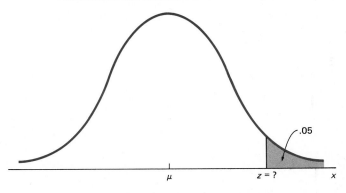

THE NORMAL CURVE AS A STATISTICAL MODEL

Now it is time to learn the purpose of theoretical distributions in general and the normal distribution in particular. *If a theoretical distribution is a close approximation to an empirical (real-world data) distribution, then we can use everything we know about the theoretical distribution to answer questions involving the empirical distribution.* For example, suppose we believe that a batch of SAT scores would be arranged within a frequency distribution to have a bell shape very similar to a distribution representing a normal distribution. We could then say that our SAT scores were approximately normally distributed and then use the standard normal distribution table to answer questions involving areas. For example, what percentage of people would have SAT scores above 600 if the mean is 500, the standard deviation is 100, and scores are close to being normally distributed? A conversion to a standard score (Z) by

$$Z = \frac{X - \mu}{\sigma}$$

would be very similar to the same *z* value from the normal curve table. So

$$Z = \frac{600 - 500}{100} = 1$$

A Z greater than 1 yields an area of 15.87%. So we answer the question "What percentage of people have SAT scores above 600?" by giving a good approximation of 15.87%. How good this approximation is depends on how close the SAT distribution is to a normal distribution.

One works problems of this type exactly as other problems involving area under the normal curve.

> *Example 8.* What percentage of people have IQ's between 90 and 110 if IQ scores are normally distributed with a mean of 100 and a standard deviation of 15? The area that we are looking for is shaded in Figure 5.12. The number of Z's above the mean is
>
> $$Z = \frac{110 - 100}{15} = .67$$
>
> The number of Z's below the mean is
>
> $$Z = \frac{90 - 100}{15} = -.67$$
>
> The area in the standard normal distribution represented by each Z is 24.86%. Thus, the target area is 24.86% + 24.86% = 49.72%. Therefore,

almost 50% of the people would have an IQ between 90 and 110.

Figure 5.12

Area under the normal curve in Example 8.

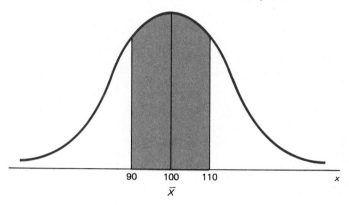

PROBABILITY AND THE NORMAL CURVE

The most important usage of theoretical distributions is in determining the probability of a single case occurring in a part of the distribution represented by a small area. Determining probabilities for single occurrences is equivalent to finding areas under curves exactly as we have been doing with the normal curve. Using the last example, suppose we ask, ''What is the probability of choosing an adult *at random* and finding that that adult has an IQ between 90 and 110. We would work the problem *exactly* as in Example 8. The probability would be .4972. (Probability is usually stated in decimals, not percentages. Therefore, we convert the area percentage into a decimal by dropping the percent sign and moving the decimal two places to the left; e.g., 49.72% = .4972.) Thus the probability of any score occurring within a specified range within a distribution is equal to the percentage of the total area encompassed by that range. We now have a way of determining probability outcomes that is different from the techniques used in the last chapter; but these probability values can be used in a fashion identical to that in which we used probability outcomes in the last chapter. *First, we specify an area within a distribution where a randomly selected score would be rare under conditions where H_0 is true. Then if we obtain a score that falls within the unlikely area, we might conclude that the score occurred in this area for some reason other than by random influences (i.e., we would reject H_0).* This rationale is the basis for the remaining statistical tests of inference that we will study in this text.

SAMPLING DISTRIBUTION OF MEANS

The particular population parameter of interest in many experiments is the mean. We can answer questions concerning the mean in exactly the same manner as we answered questions involving raw scores, provided that we have the same type of information available. For example, if IQ scores are normally distributed with $\mu = 100$ and $\sigma = 15$, we can determine the probability of selecting a person at random who has an IQ > 110. We can also determine the probability of selecting four people at random whose mean IQ > 110. To answer this type of question involving raw scores, we use the formula

$$Z = \frac{X - \mu}{\sigma}$$

To answer this type of question where means are involved, we use the formula

$$Z = \frac{\overline{X} - \text{mean of } \overline{X}\text{'s}}{\sigma_{\overline{X}}}$$

where $\sigma_{\overline{X}}$ is called the ***standard deviation of a population of means*** (or ***standard error of the mean***). Let us now examine the relationship between σ and $\sigma_{\overline{X}}$ and determine how $\sigma_{\overline{X}}$ *is computed.*

Suppose we take a population of scores whose frequency distribution is very similar to a normal distribution, such as IQ scores. This distribution has a mean of 100 and a standard deviation of 15. Now assume that we *randomly* select four scores from this group and find the mean of these four scores. Then we repeat the procedure 1000 times, so we have 1000 means, each mean representing four randomly chosen scores. If we then make a frequency distribution of these 1000 means, what will this distribution look like? It will be very close to being normally distributed. Further, the mean of this distribution of 1000 means will be very close to the mean of the population of scores from which the samples were taken. (With an infinite number of samples the mean of the sampled distribution would be the same as the mean of the parent population. Also, the sampled distribution would be exactly normally distributed.) However, the standard deviation of this distribution of means (or standard error of the mean) would be smaller than the parent population standard deviation by a factor close to $1/\sqrt{N}$. For an infinite number of samples it would be exactly $1/\sqrt{N}$. Thus for an inifinite number of samples

$$\sigma_{\overline{X}} = \frac{\sigma}{\sqrt{N}} \qquad\qquad 5.4$$

**standard error
of the mean :**

(i.e., $\sigma_{\overline{X}}$) standard deviation of a population of means

where $\sigma_{\overline{X}}$ is the population standard deviation of the population of means, σ is the population standard deviation of the set of scores from which samples are taken, and N is the number of scores in each sample mean. (*N* is *not* the number of sample means in the distribution.) If we have to estimate $\sigma_{\bar{X}}$ from a sample involving a finite number of means, the formula becomes

$$S_{\bar{X}} = S \text{ of the group of sample means} \qquad 5.5$$

where $S_{\bar{X}}$ is the estimated population standard deviation of a population of means.

What if we do not have a group of sample means with which to estimate $\sigma_{\bar{X}}$? What if we have only one sample mean? Even in this case, we can still estimate $\sigma_{\bar{X}}$, not from the *S* of a group of sample means (which cannot be done if only one sample mean exists), but from the formula

$$S_{\bar{X}} = \frac{S}{\sqrt{N}} \qquad 5.6$$

where $S_{\bar{X}}$ is the estimated population standard deviation (or standard error) of a population of means, *S* is the estimated population standard deviation from a group of sample scores, and *N* is the number of scores in the sample.

In our example involving means of IQ scores, the expected mean of the frequency distribution of means would be 100, and the expected standard deviation would be 7.5 as illustrated:

$$\sigma_{\bar{X}} = \frac{\sigma}{\sqrt{N}} = \frac{15}{\sqrt{4}} = \frac{15}{2} = 7.5$$

Notice that σ is known in this example. Suppose we wanted to estimate $\sigma_{\bar{X}}$ from the following sample of scores: 4, 6, 4, 6. We could do so by finding *S* (again see Chapter 3) for these data and *N*, and then plug these values into Equation 5.6. In this example, *S* = 1.33 and *N* = 4, so $S_{\bar{X}}$ = 1.33/$\sqrt{4}$ = .67.

Figure 5.13 illustrates how a sampling distribution of sample means would look where the number of scores in a sample (i.e., *N*) is either 4 or 16, and the standard deviation of the normally distributed population is 8. As you can see from the formula or the figure, the standard deviation of the sampling distribution of means gets smaller as the number of scores from which a sample mean is computed gets larger. A by-product of this principle is that *a sample mean will become a better estimator of a population mean as the number of scores in the sample (i.e., N) increases.*

THE CENTRAL LIMIT THEOREM

Suppose that the frequency distribution representing a population of scores from which sample means are taken is *not* normally distributed, but instead is bimodal, skewed, or has some weird shape. *The central limit theorem states that, regardless of the shape of a distribution of scores,*

Figure 5.13
Sampling distribution of means where $N = 4$ and 16.

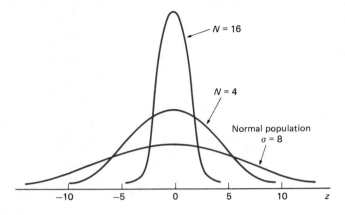

the distribution of sample means taken from that distribution (each mean consisting of N randomly selected scores) will more closely approximate a normal distribution as N gets larger. When N is infinitely large, the distribution of sample means will be normally distributed. Thus, for most empirical distributions, the sampling distribution of means taken from that distribution will be close to a normal distribution when N is above the 20–30 range. Figure 5.14 illustrates the central limit theorem.

Figure 5.14
The central limit theorem illustrated.

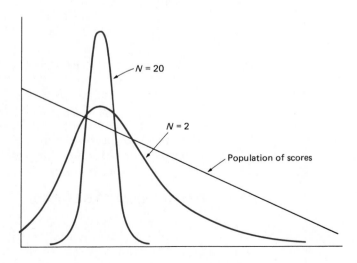

The reason that the central limit theorem works is that as N increases, the mean of a sample of scores will more closely approximate the mean of the population of scores from which that sample was taken, thus decreasing the variability of the distribution of sample means and centering this distribution around the mean of the parent distribution. This happens because when random samples of N scores are drawn from a population of scores, as N gets larger, there is a better chance that any extremely low scores in a sample will be offset by high scores, and vice versa. The net effect of these principles can be seen in Figure 5.14.

PROBABILITY AND SAMPLE MEANS

Let us go back to our original problem. "If we draw a random sample of four IQ scores and compute the mean of this sample, what is the probability that the mean of this sample will be greater than 110?" We can answer this question in the same way that we did in our other examples of area under the normal distribution. The mean of our sampling distribution of means would be expected to be 100 while the standard deviation based on $N = 4$ is 7.5:

$$\sigma_{\bar{X}} = \frac{15}{\sqrt{4}} = \frac{15}{2} = 7.5$$

See Figure 5.15. Note that σ was known and did not have to be estimated from S. Thus a Z score appropriate for our problem is

$$Z = \frac{110 - 100}{7.5} = \frac{10}{7.5} = 1.33$$

Figure 5.15
Illustrating the probability of a sample mean.

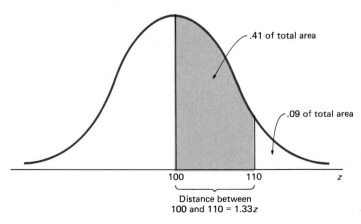

.41 of total area

.09 of total area

100   110   z

Distance between
100 and 110 = 1.33z

Looking in our normal curve table, we find that the area between μ and $1.33z$ is 40.82%, so the area beyond this point is .0918. This is the probability value for which we are looking. Thus, the probability of randomly selecting four people who would have a mean IQ greater than 110 is .0918.

THE NORMAL CURVE AND INFERENCES ABOUT EXPERIMENTS

Suppose that we did the drug-versus-placebo reaction time study described in Chapter 4 with 36 subjects. Each subject participates in both conditions as before, and, as before, we subtract each person's placebo score from his or her drug score. These subtracted scores are called *difference* (i.e., D) scores.

Suppose the mean ($\overline{D}$) of the difference scores is 2.75 and the standard deviation is 1.84. From Equation 5.6, the estimated standard deviation for a distribution of means from which our sample mean is one case is

$$S_{\overline{X}} = \frac{1.84}{\sqrt{36}} = \frac{1.84}{6} = .31$$

Now, under conditions where the null hypothesis is true (no effect for the drug in the population representing the conditions of the experiments), we would expect that $\overline{D} = 0$. However, we also recognize that in our sample the mean of the difference scores could be different from zero because of random fluctuations of a person's reaction time from one test to another. The question then becomes, ''Could a difference score mean of 2.75 have reasonably occurred by chance due to random influences?'' We can answer this question by doing our Z score transformation as before and find out within what region of the normal curve that this Z score falls. If it falls in the tail where a Z score is unlikely to occur by chance under conditions where H_0 is true, then we will reject the null hypothesis. If we use our α level criterion of .05 (see Chapter 4 to review α level), then we can mark off 5% of the area that occurs in one tail of the normal curve as an area for which a randomly selected z score (when H_0 is true) is unlikely to occur (see Figure 5.16). We mark the H_0 rejection area in one tail only because the hypothesis stated in Chapter 4 was a directional one, namely, ''Does the drug slow down reaction time?'' If we had a nondirectional hypothesis, we would split the H_0 rejection area equally between the two tails. We will elaborate on this topic in the next chapter.

From our normal curve table, we find that a z of 1.65 (rounded up) is the z value that places .05 of the area beyond this value.

Since the scores in the reference distribution appropriate for our experiment consist of means instead of raw scores, the Z transformation $Z = (X - \overline{X})/S$ becomes

α error :

(i.e., Type I error) error made by rejecting a null hypothesis that is true

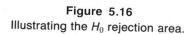

Figure 5.16
Illustrating the H_0 rejection area.

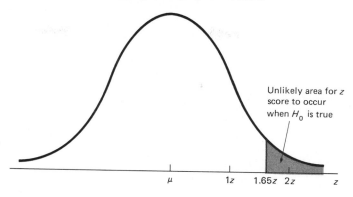

$$Z = \frac{\overline{X} - \text{mean of } \overline{X}\text{'s}}{S_{\overline{X}}}$$

where $\overline{X}$ is the mean of the difference scores in our experiment (i.e., $\overline{D}$), mean of $\overline{X}$'s is the expected mean of the $\overline{X}$'s, which is 0, and $S_{\overline{X}}$ is the estimated standard deviation of the distribution of means from which our difference score mean (i.e., $\overline{D}$) is one case. Note that in our experiment we do not know σ as we did with IQ scores. We must therefore estimate σ from the standard deviation (i.e., S) of our difference scores. (Using S to estimate σ is the typical situation in experiments.) We also estimate $\sigma_{\overline{X}}$ by using $S_{\overline{X}}$. So our Z transformation is

$$Z = \frac{\overline{X} - 0}{S_{\overline{X}}} = \frac{2.75}{.31} = 8.87$$

Since this value is within the region for rejecting H_0, we reject the null hypothesis. In our example, H_0 is population $\overline{D} = 0$.

   You may have noticed one piece of missing information in this example. In order for us to use the normal distribution as a statistical model, the frequency distribution of which our sample mean (i.e., $\overline{D}$) is one case must be normally distributed. How can we know the shape of a distribution when only one score is available? The answer involves the central limit theorem. Any sampling distribution of means taken from *any* population will be approximately normally distributed when $N > 30$ ($N = 36$ in our example).

   This Z transformation test (i.e., Z test) is appropriate only to the extent that the standard normal distribution is the appropriate theoretical distribution. It turns out that the standard normal distribution is a good approximation in this situation because N is large (that is, above 30). However, this Z test is seldom used in this situation because there is a theoretical distribution, very similar to the normal distribution, that is appropriate for both large and small samples. We will study this distribution, the *t* distribution, in the next chapter. However, it and the corresponding test, the *t* test, are used *exactly* as the test based on the normal distribution and the Z transformation that we have described.

SUMMARY

When scores are transformed, they are converted from one scale into another by a set of rules. For example, one can convert measurements listed in foot units into inch units by multiplying each foot measurement by 12. A set of scores can be transformed into standard scores (Z scores) by first subtracting the mean of the scores in the set from each score in that set and then dividing the resultant value by the standard deviation of the set of scores, i.e.,

$$Z = \frac{X - \mu}{\sigma}$$

The mean of a set of Z scores will always be 0, and the standard deviation will always be 1. A positive Z score always represents a score above the mean, and a negative Z score always represents a score below the mean.

Frequency distributions compiled from data are called *empirical* frequency distributions. Frequency distributions defined by mathematical functions are called *theoretical* frequency distributions. The *normal* distributions are a family of theoretical distributions that are (a) unimodal, (b) symmetric, (c) and have tails that get closer and closer to the *x* axis but never touch. The *standard normal distribution* is the bell-shaped distribution that results when scores that form any one normal distribution are converted into Z scores. As with any distribution converted into Z scores, the mean of the standard normal distribution is 0, and the standard deviation is 1. Standard deviation units within the standard normal distribution are called *z units*.

Z scores can themselves be transformed into other scores. *T scores* are Z scores that have been transformed into a scale having a mean of 50 and a standard deviation of 10:

$$T = 10Z + 50$$

Scores such as SAT scores are Z scores transformed into a scale having a mean of 500 and a standard deviation of 100:

$$SAT = 100Z + 500$$

Mathematicians have computed areas under the normal distribution starting from the mean (μ) and going to various *z* units and fractions thereof into a tail. The values are listed in Table B in the appendix. In solving for area in normal distribution problems, you should first draw a normal distribution and then shade in the target area.

If a theoretical distribution such as a normal distribution is a close approximation to an empirical (real-world data) distribution, then we can use knowledge about the theoretical distribution to answer questions involving the empirical distribution. Thus, we can answer questions such as, "What is the probability that a randomly selected adult would have an IQ score greater than 130?" by forming a Z score and using areas that have been marked off within the normal distribution to give a close approximation to the correct answer. Using a theoretical distribution to answer a probability question of this type is the basis for the most common types of statistical tests of inference.

If N scores are randomly selected from a population of scores, the mean of these N scores is computed, and this process is continued, the resulting frequency distribution is called a *sampling distribution of means of size N*. The mean of this sampling distribution of means would be close to the mean of the parent population from which samples were drawn. It would be exactly the same with an infinite number of sample means. The standard deviation of a sampling distribution of means with each mean comprising N scores is smaller than the standard deviation of the parent population by a factor of $1/\sqrt{N}$. Therefore,

$$\sigma_{\bar{X}} = \frac{\sigma}{\sqrt{N}}$$

The *central limit theorem* states that regardless of the shape of a distribution of scores, the distribution of sample means from this distribution, each mean consisting of N randomly selected scores, will more closely approximate a normal distribution as N gets larger. When N is infinitely large, the distribution of sample means will be normally distributed.

When N is large (i.e., above 30), the standard normal distribution can be used to find probability values from which decisions about rejecting H_0 from an experiment can be made. For example, if the transformation

$$Z = \frac{\bar{X} - 0}{S_{\bar{X}}}$$

gives a value that falls within the .05 region of the tails of a normal curve, the experimenter would reject H_0 (using $\alpha = .05$). However, in the situation described, the t distribution is usually used because it is appropriate for both large and small samples. However, it is used in exactly the same way as the normal distribution.

KEY TERMS

transformed score	Z score
standard score	z score
T score	theoretical distribution
SAT score	empirical frequency distribution
normal distributions	μ
the standard normal distribution	$\sigma_{\bar{X}}$
statistical model	$S_{\bar{X}}$
sampling distribution of means	Central limit theorem
Z test	

PROBLEMS

1. Demonstrate that the mean of a set of Z scores is 0 and the standard deviation 1 by converting the following sets of scores into Z scores and finding the mean and standard deviation of the Z scores.

	(a) 21	(b) 3	(c) 115
	25	-5	106
	19	-4	109
	13	0	110
	22	1	
		8	

2. Craig was bragging that he did better on his math exam than Brad did on his English exam. Was he correct? Assume a form of ''curved'' grading.

	Math Scores		English Scores
Craig's score	97	Brad's score	76
	93		73
	95		63
	96		61
	93		70
	90		66
			71

3. Could a person having the highest score on a test *ever* have a lower Z score than another person who had the second highest score on a different test? Explain.

4. ''You can't compare apples and oranges.'' Comment on this statement in light of this chapter.

5. As a homework assignment, a sadistic statistics instructor gives his students the task of using the computational formula to find the standard deviation of a set of 75 numbers ranging from 1500 to 1528. Which of the rules for transforming scores would greatly simplify the number-crunching (i.e., pressing keys on a calculator) task involved in finding the standard deviation? State the rule and explain why using it would make the task easier.

6. A company uses a test called the Mental Abilities Test to screen applicants for a particular training program. This test has been found to give scores that are normally distributed with a mean of 60 and a standard deviation of 15. In the past, they have admitted people having scores in the top 60% of all scores. If they again use the same criterion for admission, which of the following applicants are likely to be admitted? Pam (score 71), Lois (score 45), Charles (score 48), Kathy (score 59). Outline the procedure used to make the decision.

7. Convert each person's score in problem 6 into (a) T scores; (b) an SAT scale score.

8. Given the following information, fill in the Z scores, T scores, and SAT scale scores:

X	μ	σ	Z Score	T Score	SAT Scale Score
31	35	10			
73	80	5			
50	25	10			

9. Will is about to have his driving privileges suspended because his father has noticed a 40 marked on his exam. Quick-thinking Will responds by saying that the 40 is not his real score but his T score, and that his score was actually the highest of all people taking the exam. Could Will have been telling the truth?

10. Assume IQ scores are normally distributed with $\mu = 100$ and $\sigma = 15$. What percent of people have IQ scores

    a   above 85

    b   below 70

    c   above 130

    d   between 115 and 130

    e   between 87 and 118

11. What is the probability of a randomly selected person having an IQ ($\mu = 100$; $\sigma = 15$)

    a   below 120

    b   above 120

    c   less than 70 *or* greater than 130 (see Chapter 4)

    d   between 85 and 90 or greater than 100

    e   between 90 and 100 or less than 110 (this is not mutually exclusive; sketch target area first)

    f   above 100, with a randomly selected second person having an IQ below 100

12. Above what T value will 5% of T scores fall?

13. What percentage of people have SAT scores between 400 and 600?

14. Compute $S_{\bar{X}}$ for each of the following sets of raw scores:

a	b	c
17	5	3
15	1	1
9	9	11
13	8	9
	0	15
	6	23
		40

15. From the random number table (Table A in the appendix) choose three random numbers (e.g., the first three numbers in some column) and compute the mean of those three numbers. Record the number representing the mean and repeat the procedure using the next three numbers in the column chosen. Record that value and repeat the procedure until 12 means have been found.

    a   Using the formula $S_{\bar{X}} = S/\sqrt{N}$, find $S_{\bar{X}}$ for each of the 12 sample means that you selected. Notice that, since $S_{\bar{X}}$ is an estimate, you will likely get different values of $S_{\bar{X}}$ for different samples.

    b   Now compute the S of the 12 sample means themselves. This value is also $S_{\bar{X}}$, that is, an estimate of $\sigma_{\bar{X}}$.

    c   In this case, we can compute σ and $\sigma_{\bar{X}}$ exactly and not have to estimate from samples. Since each random number will occur equally often in the long run (i.e., infinity), use the numbers 0, 1, 2, 3, 4, 5, 6, 7, 8, 9 to find σ. Divide σ by $\sqrt{3}$ ($N = 3$ in our example). This value is $\sigma_{\bar{X}}$. Notice the closeness of this value and estimates of this value in (a) and (b).

    d   Round each of the means to the nearest whole number; then make a frequency polygon (Chapter 2) of these numbers. Scale the x axis from 0 to 9. Around what values on the x axis does the frequency polygon center? Why?

16. Find $S_{\bar{X}}$:

    a $S = 39, N = 12$

    b $S = 400, N = 100$

    c $S = 50, N = 16$

    d $S = 1, N = 25$

17. If the mean of a distribution of means ($N = 8$) is 50 and $S_{\bar{X}}$ is 4, what is the probability that eight randomly drawn scores will have an $\bar{X}$

    a greater than 60

    b less than 54

    c greater than 56 or less than 44

18. Use the Z test described in the last section to determine if a difference score mean (i.e., $\bar{D}$) of 2.46 is different from zero (i.e., can H_0 be rejected?); $S = 3.1$, $N = 40$, and $\alpha = .05$, directional test.

19. Do the same test as in problem 18 with

    a $S = 1.6, N = 50, \sigma = .05$, directional test, $\bar{D} = 3.5$

    b $S = 8.3, N = 50, \sigma = .05$, directional test, $\bar{D} = 3.5$

    c $S = 6.1, N = 40, \sigma = .05$, nondirectional test, $\bar{D} = 2.3$. [*Hint:* The rejection region has half of the rejection area in each tail.]

The *t* Test

Note to the Student

Many of the concepts presented in Chapters 3, 4, and 5 are used to develop the material in this chapter. A review of Chapters 3, 4, and 5 before beginning this chapter will allow you to understand this chapter more easily. Specifically, you will need to know the following terms and concepts. Chapter 3: $\bar{X}$, SS, S^2, S, σ, and σ^2. Chapter 4: H_0, α level, directional test, nondirectional test, and sign test. Chapter 5: transformed scores, z score, $S_{\bar{X}}$, Z test, normal distribution, probability and areas under curves, and statistical model.

THE LOGIC OF THE *t* TEST

We will now show how a *t* test might be used to analyze our drug example in Chapter 4, where we were trying to determine whether a drug slowed down a person's reaction time. Observe that, in this example, the sample size (N) is too small for the Z test described in Chapter 5 to be appropriate. However, as noted in Chapter 5, the *t* test can be used for both large and small samples. To begin, let us suppose that we had the same design as before. Each of 11 people has taken both the drug and a placebo on different occasions and has had reaction times measured on each occasion. As before, we will assume no practice or carryover effects. Now what would we expect under conditions of the null hypothesis (i.e., the drug has no effect on reaction time)? As with the sign test, we would expect a random distribution of pluses and minuses when we subtract a person's score in one condition from his or her score in the other condition. As described in Chapter 4, scores obtained in this manner are called difference scores (i.e., D scores). Although not important for the sign test, the magnitude of the difference scores should randomly vary such that, when averaged together, these difference scores would be expected to have a mean of zero (i.e., $\overline{D} = 0$). *The term **expected** refers to the value that would be obtained with an infinitely large number of scores. However, we don't have an infinitely large number of D scores; we only have 11. **Therefore, because of random or chance factors (also called sampling error), the mean of the difference scores (i.e., $\overline{D}$) from our experiment would likely be close to zero (assuming H_0 to be true), but not necessarily exactly zero.***

A coin-toss analogy may help you to better understand the relationship between expected values and values obtained from samples. If you have 11 coins and drop them on the floor, how many heads would you expect? The most likely numbers are 5 and 6, but getting four or seven heads would also be common, and getting three or eight heads would not be surprising. As we saw with the sign test, we had to go out to 9 to find an event (based on $N = 11$) that we would consider to be unlikely (based on $\alpha = .05$).

Now we may ask exactly the same kind of question with an obtained difference score mean ($\overline{D}$) as we asked with the sign test and the Z test. Based on our drug example, how likely is it to get a difference score mean of a particular size if H_0 were really true? Intuitively, you might think that a value of, say, .37 would be likely, whereas a value of, say, 5.9 would be very unlikely. But where is the cutoff? At what point do we say values up to this size are values that we will assume occurred with H_0 being true and values beyond this point are too unlikely for us to accept H_0 as true? We could easily answer this question if we knew the appropriate frequency distribution for our example.

simulation:

an attempt to replicate something, usually under simplified conditions

Just as with coins, we could pretty accurately estimate what this distribution would be by doing a *simulation*. With this simulation we could repeat the experiment 100 times with placebos (thus ensuring that H_0 is true) for both conditions. Using $\alpha = .05$, we could then determine the 5 most extreme positive (reflecting the directional test) difference score means out of the 100 we obtained. (Each mean would be derived by first subtracting each subject's score in one condition from that subject's score in the other condition, to yield a difference score, and then adding the difference scores for all subjects together and dividing by the number of difference scores, which in our example is 11; i.e., $\overline{D} = \Sigma D/N$.) We would use the value of the fifth most extreme positive mean as our cutoff. If the mean that we obtained in our real experiment were greater than this cutoff value, we would say that a value of this size is too large for us to accept as a chance occurrence, so we reject H_0.

Obviously, finding a cutoff value for H_0 through simulation is laborious. Fortunately there is an easier way. This easier way is the t test, which is identical to the Z test from Chapter 5, except that a slightly different cutoff value is used. Instead of the normal distribution, the t test is based on a family of theoretical distributions called the t *distributions*. These distributions are very similar in shape to the standard normal distribution. *A t distribution is more appropriate than the standard normal distribution to use as a statistical model when population Variances are unknown and have to be estimated from samples.* Since this is almost always the case with empirical research, a t distribution is almost always preferred over the standard normal distribution as a statistical model for empirical studies that compare data from two conditions. In the one situation (large N) where the standard normal distribution is also an appropriate statistical model, a t distribution and

computed t:

value obtained from t in a t test formula

The t test uses a t distribution to provide a cutoff value that is used to determine whether to accept a sample mean (or difference between means) as a chance occurrence. This is done by first computing a t value in the same way as a Z score in a Z test. The *computed t value used in a t test is a transformed score just as the small Z value used in a Z test was a transformed score.* The t transformation involves converting a mean or difference between means into a t value by using the format for standard scores that we studied in Chapter 5. We will use a t value with respect to the t theoretical distribution exactly as we used a Z value with respect to the standard normal distribution. As with the Z test, we want to know the probability of getting a difference between means (i.e., $\bar{X}_1 - \bar{X}_2$ or $\bar{D}$) of the obtained size if the population mean difference is zero and the estimated population standard deviation is $S_{\bar{D}}$ (i.e., how likely is it to obtain a t value of the obtained size if H_0 is true?). The formula for the t value for dependent groups is the obtained mean minus the mean expected when H_0 is true, divided by the standard error of the mean of the difference scores. Thus:

$$t = \frac{\bar{D} - \bar{D}_{\text{expected}}}{S_{\bar{D}}}$$

$$t = \frac{\bar{D}}{S_{\bar{D}}} \qquad (\text{since } \bar{D}_{\text{expected}} = 0) \qquad\qquad 6.1$$

The value obtained from the computed t will tell us how far away it is from the most likely t value when H_0 is true. (This most likely t value is, of course, zero.) We then compare the computed t value with the cutoff t value that is appropriate for a particular sample size and α level. (For α level review, see Chapter 4.) *If the computed t value is*

critical t:

the value that separates a t distribution into parts such that the tail or tails contain the area designated by the chosen α level; if the computed t exceeds the critical t, H_0 is rejected.

larger, it implies that a mean of difference scores this large is not a likely occurrence with H_0 being true, so we would reject H_0. If the computed t value were smaller than the cutoff or critical t, we would retain H_0 (see Figure 6.1). This critical value is based on $N = 11$ for a directional test with $\alpha = .05$.

The critical t value will vary as a function of the α level chosen. For example, had we used $\alpha = .01$ instead of .05, only 1% of the area would be in the tail and we would have to go out 2.76 t units from the center of the distribution to find the cutoff value. Note also that in this example the entire rejection region for H_0 falls in one end of the distribution. This occurs because our hypothesis is a directional one (again see Chapter 4).

Figure 6.1
A *t* Distribution

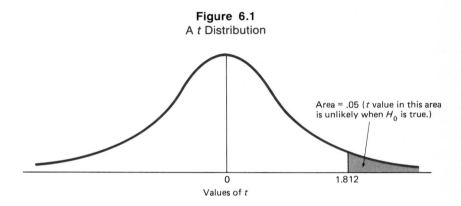

Area = .05 (*t* value in this area is unlikely when H_0 is true.)

0

1.812

Values of *t*

Now where did the critical t come from? It came from mathematicians dividing one of the t distributions into a 95%–5% split and marking that split along the x axis in terms of units of t. Unlike the standard normal distribution, the t distributions vary somewhat in shape as a function of the number of scores in a sample. The t distribution in Figure 6.1 is the one appropriate for a sample size of 11. The critical t value came from Table C in the appendix, which lists various critical t values for commonly used α levels.

The practical importance of the t distributions is that, given a particular sample size, one t distribution will be a close approximation to the distribution of sample means (having the same sample size) that we would obtain through thousands of replications of an experiment (such as our drug experiment) under conditions where H_0 is true. Thus, as previously described, if given a set of difference score means from an experiment and an estimate of how such means would be expected to vary from zero when H_0 is true, we can perform a transformation (i.e., $t = \overline{D}/S_{\overline{D}}$) of our obtained difference score mean ($\overline{D}$) and compare that value to a critical t value from an appropriate t distribution. In our example, if we go 1.812 t units from the midpoint of a t distribution, based on $N = 11$, we partition the t curve into one tail containing .05 of the area (which is appropriate for a directional test with $\alpha = .05$) and the rest of the curve containing .95 of the area. Again see Figure 6.1. Thus, if a computed t value is greater than 1.812, it would be considered to be too large to have occurred under conditions where H_0 is true (using $\alpha = .05$). Therefore, for t values larger than 1.812, H_0 would be rejected. The process of computing a t value and comparing it to a critical t value from a table in order to reject or retain H_0 is called *doing a t test*.

THE *t* DISTRIBUTIONS

So why isn't the t table of the same format as the normal curve table? It could be, but the problem is that there is a slightly different t curve for every sample size. If the t tables were printed as the normal curve table, half of the book would be t tables. Therefore, only the important cutoff values are reproduced in the t table. See Tables B and C in the appendix.

As can be seen from Figure 6.2, the t distributions are very similar to the normal curve. The most important of these similarities are that both

(a)   Are symmetrical
(b)   Are bell-shaped
(c)   Have infinitely long tails
(d)   Have an area equivalent to a probability of 1

The major difference is that the t distribution is flatter than the standard normal distribution and has more area in the tails. The degree of flatness depends on the sample size or its closely related concept, the *degrees of freedom*. A memory aid to help you remember this relationship is to think of the normal curve as a wire under your feet. With infinite degrees of freedom, the t distribution and the normal curve are the same. As the sample size gets smaller, you place your foot harder on the wire, flattening it out more, and spreading out the ends.

DEGREES OF FREEDOM

The *degrees of freedom* (df) *associated with a set of scores refers to the number of scores within the set that are free to vary under some constraint on the set.* For example, suppose that a set of four scores has the constraint that their sum must be 20. As soon as three scores are deter-

Figure 6.2
Comparison of the normal curve and *t* distributions for *df* = 5 and *df* = 15.

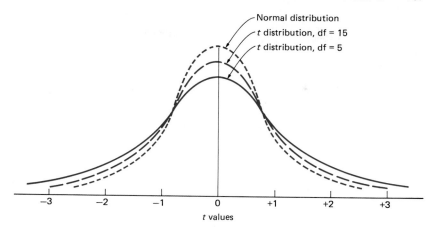

degrees of freedom:

the number of scores within a set that are free to vary given some constraint or constraints on the set

mined, the fourth is automatically determined by the constraint. For example, if three scores are 1, 9, 3, the fourth score must be 7. In statistical usage, degrees of freedom usually involve the number of scores in a group minus 1, or $N - 1$. The reason for the loss of a df within a group is that each group is used to make a Variance estimate (i.e., S^2) of the population Variance. The constraint creating the loss of a df comes from the fact that the deviation scores [i.e., $\Sigma (X - \bar{X})$] used in the formula for S^2 must sum to zero.

With respect to the *t* distribution, degrees of freedom are based on the number of scores in a sample minus 1. Since we had 11 difference scores, we have 10 df in this situation. The critical *t* values in Table C are listed by degrees of freedom down the side and by α levels across the top. Thus, in our example, we had 10 df with an α level of .05 and a directional test. The critical *t* was therefore 1.812. You should check this for yourself.

DIRECTIONAL VERSUS NONDIRECTIONAL *t* TESTS

Any hypothesis can be stated such that either a directional (also called one-tailed) or nondirectional (also called two-tailed) test is appropriate. If a nondirectional (two-tailed) test is used with $\alpha = .05$, the 5% rejection region is divided such that .025 of the area is in each tail, as shown in Figure 6.3. Because the rejection area with the nondirectional test is divided between the two tails, the absolute value (value disregarding the sign of the number) of the *t* value representing the cutoff point *will always be larger for the nondirectional test*. Effectively, this means that it will be more difficult to reject H_0 with a nondirectional test than with a directional test.

Figure 6.3

Comparison of directional and nondirectional *t* tests for 10 degrees of freedom and $\alpha = .05$.

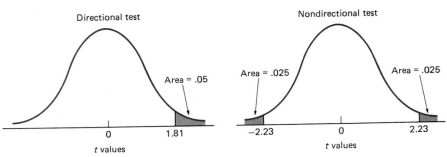

Notice that since the area in a tail is split into two equal parts with a nondirectional test, the cutoff value for a nondirectional test that has a particular α level will always be the same as the cutoff value for a directional test for an α level that is twice as large. For example, 2.228 for 10 df is the $\alpha = .05$ cutoff for a nondirectional test and also the $\alpha = .025$ cutoff for a directional test. This relationship is utilized in the headings of the *t* table in the back of the book.

General Procedure for Doing *t* Tests

(a)   Determine directionality, degrees of freedom, and α level to be used in test.

(b)   Based on step 1, find the cutoff (i.e., critical) *t* value from a *t* table.

(c)   Based on data from your experiment, compute a *t* value and disregard the sign.

(d)   Compare your computed *t* value with the cutoff *t*. If it is greater than the cutoff *t*, reject H_0; if less, retain H_0. For *t* tests, H_0 is stated as $H_0: \mu_1 = \mu_2$. (The population means of the two conditions are equal.) *For directional tests, the computed t must be larger than the critical t, and the difference between the means of the two conditions must be in the **predicted direction** for H_0 to be rejected.*

COMPUTATIONAL OUTLINE FOR DOING DEPENDENT GROUPS *t* TESTS

repeated-measures design:

design wherein each subject participates in every condition of the independent variable

An experimental design may require that different groups of subjects participate in each condition of the experiment (between-subject designs). (The analyses for this type of design will follow the current discussion.) Alternatively, the same subjects may participate in all of the conditions of the experiment (repeated-measures designs), or groups may consist of members who are matched on relevant vari-

matched-groups design:

a design wherein each subject in one condition has a matched subject in the other conditions; matching is done with respect to the dependent variable

ables (matched-group designs). *The statistical analyses for repeated-measures designs and matched-group designs are identical,* so both will be subsumed under the heading of *dependent groups tests.*

To illustrate the procedure for doing a dependent groups *t* test, we list the data from the drug-versus-placebo reaction time experiment in Chapter 4 in Table 6.1. The denominator of the computed *t* value will be the standard error of the mean (see Chapter 5) using the deviation scores.

TABLE 6.1
Reaction Time Scores for Drug-versus-Placebo Experiment

Subject	Placebo	Drug	D = Drug − Placebo
1	30	28	−2
2	29	33	+4
3	31	34	+3
4	23	25	+2
5	29	31	+2
6	33	32	−1
7	28	30	+2
8	31	35	+4
9	33	36	+3
10	29	33	+4
11	28	34	+6

Following the first two steps in the general procedure for *t* tests, we have a one-tailed test, with 10 df, and we will use the typical α level of .05. The df is the number of difference scores minus 1 ($N - 1$). The cutoff *t* value from Table C in the appendix is therefore 1.812. You should check this for yourself.

For step 3, we need to compute *t*, where

$$t = \frac{\bar{D}}{S_{\bar{D}}}$$

6.1

(*t* equals the mean of the *D* scores divided by the standard error of the mean of the *D* scores, i.e., $S_{\bar{D}}$). From Table 6.1, $\bar{D} = 2.45$. Putting $S_{\bar{D}}$ into computational form, we have

(definitional formula)   (computational formula)

$$S_{\bar{D}} = \frac{S_{\bar{D}}}{\sqrt{N}} = \sqrt{\frac{\Sigma(D - \bar{D})^2}{N(N-1)}} = \sqrt{\frac{N\Sigma D^2 - (\Sigma D)^2}{N^2(N-1)}}$$

From the data in our experiment, we have

(a) $\Sigma D = -2 + 4 + 3 + \cdots + 6 = 27$
(b) $(\Sigma D)^2 = 27 \times 27 = 729$
(c) $\Sigma D^2 = (-2)^2 + (4)^2 + (3)^2 + \cdots + (6)^2 = 119$
(d) $N = 11$
(e) $\bar{D} = 2.45$

Plugging these values into the computational formula gives

$$t = \frac{2.45}{\sqrt{\dfrac{(11 \cdot 119) - 729}{(11 \cdot 11)(10)}}} = 3.54$$

Since $3.54 > 1.81$, we reject H_0 and infer that the drug slowed down reaction time. A typical way in which these results would be presented in a journal article would be the following: "A dependent groups t test showed the difference between the means of the drug and placebo groups to be significant, $t(10) = 3.54$, $p < .05$, with the drug group having slower reaction times." In this statement, *significant* means that H_0 was rejected, and the value in parentheses refers to the degrees of freedom in the critical t. The p represents the probability of the computed t (e.g., 3.54) occurring if H_0 were true. The last value, .05, represents an α level.

   Although we came to the same conclusion with this test as with the sign test, this is not always the case. If the scores that make up the data are reasonably normally distributed, a parametric test (e.g., the t test) will be more *powerful* (i.e., it has a better chance of correctly rejecting H_0) than a nonparametric test (e.g., the sign test). However, with the same rationale, if a nonparametric test is significant, then a parametric test on the same data will also be significant. It is the converse that is not true, which is why parametric tests should be used when the major assumptions for their use (i.e., normally distributed scores reflecting an interval or ratio scale) are met.

power:

the probability of correctly rejecting H_0 in a statistical test

THE INDEPENDENT GROUPS *t* TEST

independent groups designs:

designs wherein different groups of subjects participate in each condition of the independent variable

Consider a situation where we have performed an experiment that has two conditions with different groups of subjects in each condition. *Because each subject participates in only one condition (not both as with a repeated-measures design) and because no matching of subjects is involved, this type of design is called an* **independent groups design.** As before with the dependent groups design, we wish to see if the population mean of one group is different from the population mean of the other group (i.e., can we reject H_0?). For example, suppose we had performed our drug-versus-placebo reaction time experiment using different subjects in each of the conditions. We still want the same question answered, namely, are the means of the two groups sufficiently different so that we may conclude that this difference did not happen because of random influences (i.e., sampling error)? We answer the question by doing a *t* test as we did with the dependent groups. The difference between the two types of *t* tests is in the term reflecting variability in the denominator. This term is now the ***standard error of the difference between means for independent groups,*** or $S_{\bar{X}_1 - \bar{X}_2}$ instead of $S_{\bar{D}}$ (i.e., the standard error of the mean for the difference scores).

standard error of the difference between means:

the standard deviation of a set of difference between mean (i.e., $\bar{X}_1 - \bar{X}_2$) scores

Exactly what is $S_{\bar{X}_1 - \bar{X}_2}$? It is the *estimated* population standard deviation of a set of scores that comprise differences between sample means (i.e., subtracting one mean from another) if each sample mean were taken from the same normally distributed population (or from two populations having the same μ and σ). To illustrate, if we gave two randomly selected groups (of size N) placebos, found the mean for each group, then subtracted one mean from the other, we would have a score representing the difference between these means. If we repeated this procedure several times, we would have a set of scores representing differences between means. The standard deviation of this set of scores is $S_{\bar{X}_1 - \bar{X}_2}$, which is an estimator of the population standard deviation $\sigma_{\bar{X}_1 - \bar{X}_2}$. However, analogous to our discussion of $S_{\bar{X}}$ in Chapter 5, we usually have to estimate $\sigma_{\bar{X}_1 - \bar{X}_2}$ from data involving only one score representing $\bar{X}_1 - \bar{X}_2$. We will shortly show how to do this.

The mean of this (or any) sampling distribution of differences between sample means would be expected to be zero. The reason is that sometimes subtracting one sample mean from another will yield positive numbers and sometimes it will yield negative numbers. In the long run (i.e., infinity), the positive and negative numbers will sum to the same value, yielding a mean of zero. However, any value of $\bar{X}_1 - \bar{X}_2$ can differ markedly from zero. Knowing the standard deviation of $\bar{X}_1 - \bar{X}_2$ (i.e., $S_{\bar{X}_1 - \bar{X}_2}$) will allow us to determine a range of values for $\bar{X}_1 - \bar{X}_2$ that would occur a certain percentage of the time when H_0 is true (i.e., when $\mu_1 = \mu_2$). If a value for $\bar{X}_1 - \bar{X}_2$ were beyond a certain cutoff, we would reject H_0. A simulation may help you to better understand this.

Suppose that we give placebos to two randomly selected groups of size N. We then find the mean reaction time for each group and subtract one mean from the other. If we repeated this experiment, say, 100 times, we would have a pretty good idea of how large a difference between means would have to be in order to be considered an unlikely occurrence when H_0 is true. Of course, H_0 would be true in this situation (e.g., both groups were given placebos). See Figure 6.4.

Figure 6.4

Sampling distribution of differences between means for 100 cases, hypothetical data.

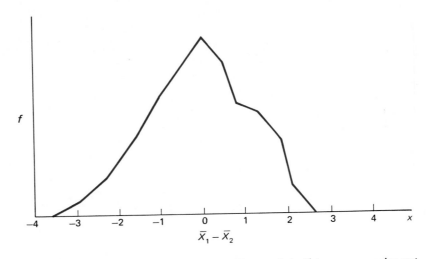

Based on the situation depicted in Figure 6.4, if in an experiment we found a difference greater than $+4$ or less than -4 when we subtracted one mean from another we could feel pretty confident that this would be a very unlikely event to have occurred if H_0 were really true. Therefore, if our difference between the drug and placebo groups were, say, 4.69, we would reject H_0 and infer that the difference between the drug and placebo groups was due to the drug affecting reaction time.

However, as with our dependent groups t test, we can estimate what a distribution such as the one in Figure 6.4 would look like and, more importantly, where the cutoff values would be by using the appropriate t distribution as a statistical model. Once we know the appropriate t distribution we transform $\bar{X}_1 - \bar{X}_2$ into a t value by using the formula

$$t = \frac{\bar{X}_1 - \bar{X}_2}{S_{\bar{X}_1 - \bar{X}_2}}$$

6.2

If our computed t value fell far enough in one of the tails of the t distribution (past the value determined by our α level), this would be an unlikely event to have occurred with H_0 being true, so we would reject H_0. The appropriate t distribution for a t test is determined from the number of degrees of freedom involved in both groups:

$$\text{df}_{\text{group 1}} = N_1 - 1$$

$$\text{df}_{\text{group 2}} = N_2 - 1$$

$$\text{df}_{\text{groups 1, 2}} = N_1 - 1 + N_2 - 1 = N_1 + N_2 - 2$$

In our example, we had 11 subjects in each group, so total df = (11 − 1) + (11 − 1) = 20. The appropriate *t* distribution for our example is therefore the one with 20 df. Based on a particular α level, we can therefore find our cutoff value from the *t* table for the *t* distribution having 20 df.

To summarize our discussion, we first calculate a *t* value from our data and then see if that *t* value is large enough to fall within the rejection region of the appropriate *t* distribution (i.e., our computed *t* value is larger than the cutoff *t*). If our computed *t* is larger than our critical *t*, we reject H_0. If our computed *t* is less than our critical *t*, we retain H_0. This process is called doing a *t* test. There are two types of *t* tests: dependent groups and independent groups. *The first step in doing any t test is to determine, based on the experimental design, which test is appropriate.*

CALCULATING THE INDEPENDENT GROUPS *t* VALUE WHEN $N_1 = N_2$

When $N_1 = N_2$ (i.e., the number of scores in each condition is the same), then our estimate of $\sigma_{\bar{X}_1 - \bar{X}_2}$ (i.e., $S_{\bar{X}_1 - \bar{X}_2}$) is obtained from the variances within each condition, as shown in Equation 6.3:

$$S_{\bar{X}_1 - \bar{X}_2} = \sqrt{\frac{S_1^2}{N_1} + \frac{S_2^2}{N_2}} \qquad 6.3$$

Therefore, to transform $\bar{X}_1 - \bar{X}_2$ into a *t* value, we use the formula

$$t = \frac{\bar{X}_1 - \bar{X}_2}{\sqrt{\frac{S_1^2}{N_1} + \frac{S_2^2}{N_2}}} \qquad 6.4$$

where $\bar{X}_1$ and $\bar{X}_2$ represent the means of the two conditions from our experiment, S_1^2 and S_2^2 represent the Variances of the two conditions, and N_1 and N_2 represent the sample sizes of the two conditions.

When $N_1 \neq N_2$

When $N_1 \neq N_2$, a better estimate of $\sigma_{\bar{X}_1 - \bar{X}_2}$ can be obtained by substituting S_{pooled}^2 for S_1^2 and S_2^2, giving

$$t = \frac{\bar{X}_1 - \bar{X}_2}{S_{\bar{X}_1 - \bar{X}_2}} = \frac{\bar{X}_1 - \bar{X}_2}{\sqrt{\frac{S_{pooled}^2}{N_1} + \frac{S_{pooled}^2}{N_2}}}$$

$$= \frac{\bar{X}_1 - \bar{X}_2}{\sqrt{S_{pooled}^2 \left[\frac{1}{N_1} + \frac{1}{N_2} \right]}}$$

The term S^2_{pooled} is the *pooled Variance* of the two conditions. It is the weighted average of S^2_1 (Variance of the first group) with S^2_2 (Variance of the second group). By *weighted* we mean that different N's that may be in each group are accounted for in the average. The formula for computing S^2_{pooled} is

$$S^2_{pooled} = \frac{SS_1 + SS_2}{N_1 + N_2 - 2} \qquad 6.6$$

Therefore, inserting Equation 6.6 into Equation 6.5, we get

$$t = \frac{\bar{X}_1 - \bar{X}_2}{\sqrt{\dfrac{SS_1 + SS_2}{N_1 + N_2 - 2}\left[\dfrac{1}{N_1} + \dfrac{1}{N_2}\right]}} \qquad 6.7$$

Example 1. To find a computed t value using these formulas, we need the following values:

(a) $\bar{X}_1 - \bar{X}_2 =$
(b) $N_1 =$
(c) $N_2 =$
(d) $SS_1 =$
(e) $SS_2 =$

Suppose that we did our drug-versus-placebo experiment with four different people in each group (an N of 4 is too small in this situation for reliable results, but we can more easily demonstrate the calculations)

Placebo (X_1)	Drug (X_2)
31	32
33	36
28	28
27	29
$\bar{X}_1 = 29.75$	$\bar{X}_2 = 31.25$

We begin by finding ΣX and ΣX^2 for each group of scores:

Placebo $\Sigma X_1 = 31 + 33 + 28 + 27 = 119$

$\Sigma X_1^2 = 31^2 + 33^2 + 28^2 + 27^2 = 3563$

Drug $\Sigma X_2 = 32 + 36 + 28 + 29 = 125$

$\Sigma X_2^2 = 32^2 + 36^2 + 28^2 + 29^2 = 3945$

Therefore, listing the necessary values and plugging into Equation 6.7, we have

(a) $\bar{X}_1 - \bar{X}_2 = 29.75 - 31.25 = -1.50$
(b) $N_1 = 4$
(c) $N_2 = 4$

(d) $SS_1 = \dfrac{4(3563) - (119)^2}{4} = 22.74$

(e) $SS_2 = \dfrac{4(3945) - (125)^2}{4} = 38.76$

(f) $S_{\bar{X}_1 - \bar{X}_2} = \sqrt{\dfrac{22.74 + 38.76}{4 + 4 - 2}\left(\dfrac{1}{4} + \dfrac{1}{4}\right)}$

$\qquad = 2.26$

Now

$$t = \frac{\bar{X}_1 - \bar{X}_2}{S_{\bar{X}_1 - \bar{X}_2}} = \frac{-1.50}{2.26} = -.66$$

Since $N_1 = N_2$, we should get the same answer by using the simpler version of $S_{\bar{X}_1 - \bar{X}_2}$. We have

$$S_1^2 = \frac{SS_1}{N_1 - 1} = \frac{22.74}{3} = 7.58$$

$$S_2^2 = \frac{SS_2}{N_2 - 1} = \frac{38.76}{3} = 12.92$$

so by Equation 6.4,

$$t = \frac{29.75 - 31.25}{\sqrt{\dfrac{7.58}{4} + \dfrac{12.91}{4}}} = \frac{-1.50}{\sqrt{5.125}} = \frac{-1.50}{2.26} = -.66$$

As advertised, both techniques provide the same answer. *Once we establish the directionality of the test, we can ignore the sign of both t and the critical t and use absolute values.*

The critical t for $\alpha = .05$, 6 df, and a directional test (step 1) is 1.943 (step 2). The appropriate df used in finding a critical t value from Table C is $N_1 - 1 + N_2 - 1 = N_1 + N_2 - 2$. *Note that df are computed differently for independent groups designs compared with dependent group designs.*

Since .66 (step 3) < 1.943, retain H_0 (step 4), where H_0 is $\mu_1 = \mu_2$. Observe that the directional test asks if the drug slows down reaction time. Should the results show that the mean reaction time for the drug condition was *faster* than for the placebo conditions, H_0 is automatically retained with no statistical test being necessary. This situation can occur only for a directional test.

Example 2. From the following experimental situation, let us *first determine which type of t test (dependent groups or independent groups) is appropriate,* then do it. A school administrator receives two sets of programmed learning materials covering the same topic. She wishes to find out if one set is superior to the other in terms of student mastery of the material. If one set is superior, she will strongly recommend that that set be used by all teachers teaching the relevant material. If not, the teachers may choose either one. Ms. Wilson volunteers her 20 students for an experiment to decide the issue. The students in her class are given either set A or set B on a random basis. At the end of a specified time, a 25-point mastery test is given. The administrator sets the criterion for rejecting H_0 at $\alpha = .05$. Do the appropriate test given the following data.

Set A Scores	Set B Scores
24	21
19	23
21	11
23	17
17	15
13	19
22	22
19	13
23	15
20	12

First, we decide whether to do an independent or dependent groups t test. Since 20 students were randomly placed into two groups, there is no matching of students, nor did students take both sets of learning materials. Therefore, the *independent* groups t test is appropriate.

Step 1. The α level was chosen to be .05. Since there is interest in either set A or set B being better, the appropriate test is *nondirectional*. There is $N_1 - 1 + N_2 - 1 = 9 + 9 = 18$df in the appropriate t distribution.

Step 2. Based on a nondirectional test with 18 df and $\alpha = .05$, the critical t value is 2.101.

Step 3. Since $N_1 = N_2$, we can use Equation 6.4, the shorter version of the independent groups t test, to compute the t value:

$$t = \frac{\bar{X}_1 - \bar{X}_2}{\sqrt{\dfrac{S_1^2}{N_1} + \dfrac{S_2^2}{N_2}}}$$

The values for the equation are

(a) $\Sigma X_1 = 24 + 19 + 21 + 23 + 17 + 13 + 22 + 19 + 23 + 20 = 201$

(b) $N_1 = 10$

(c) $\bar{X}_1 = \Sigma X / N_1 = 201/10 = 20.1$

(d) $\Sigma X_2 = 21 + 23 + 11 + 17 + 15 + 19 + 22 + 13 + 15 + 12 = 168$

(e) $N_2 = 10$

(f) $\bar{X}_2 = 16.8$

(g) $\Sigma X_1^2 = 24^2 + 19^2 + 21^2 + 23^2 + 17^2 + 13^2 + 22^2 + 19^2 + 23^2 + 20^2$
$= 4139$

$(\Sigma X_1)^2 = (201)^2 = 40{,}401$

$$S_1^2 = \frac{N_1 \Sigma X_1^2 - (\Sigma X_1)^2}{N_1(N_1 - 1)} = \frac{10(4139) - 40{,}401}{(10)(9)} = 10.99$$

(h) $\Sigma X_2^2 = 21^2 + 23^2 + 11^2 + 17^2 + 15^2 + 19^2 + 22^2 + 13^2 + 15^2 + 12^2$
$= 2988$

$(\Sigma X_2)^2 = (168)^2 = 28{,}224$

$$S_2^2 = \frac{N_2 \Sigma X_2^2 - (\Sigma X_2)^2}{N_2(N_2 - 1)} = \frac{10(2988) - 28{,}224}{(10)(9)} = 18.4$$

Therefore,

$$t = \frac{20.1 - 16.8}{\sqrt{\dfrac{10.99}{10} + \dfrac{18.4}{10}}} = \frac{3.3}{\sqrt{2.94}} = 1.92$$

Step 4. $1.92 < 2.101$; therefore retain H_0.

There is no evidence from the preceding statistical test that one set of materials is superior to another.

Example 3. An investigator wishes to examine a certain type of amnesia patient to see if short-term memory span is also affected. He has available six patients who have this amnesia diagnosis. He therefore matches each patient with a nonamnesic person in terms of age, background, and cognitive abilities apart from those involved in the amnesic syndrome. He then gives each of the 12 subjects a short-term memory span test for strings of digits. The investigator uses $\alpha =$ .05 and a nondirectional test. (A case could also be made for a directional test in this situation, but most investigators prefer the more conservative, less controversial nondirectional test.) The data are listed in the table. Do the appropriate test.

Pair	Normal	Amnesic
1	7	6
2	9	7
3	8	7
4	7	5
5	6	6
6	8	7

From the design of the study, we determine that a dependent groups *t* test (Equation 6.1) is appropriate. We therefore subtract each amnesia score from the paired normal counterpart to obtain difference (D) scores. Using our four-step outline, we have

Step 1. $\alpha = $ .05, nondirectional test with $N - 1$ (N equals the number of D scores) for 5 df.

Step 2.   Based on the information in step 1, the critical *t* is 2.571.

Step 3.   Computing the *t* value using Equation 6.1, we have

$$t = \frac{\bar{D}}{S_{\bar{D}}}$$

(a)   The D values are 1, 2, 1, 2, 0, 1
(b) $\Sigma D = 1 + 2 + 1 + 2 + 0 + 1 = 7$
(c) $\Sigma D^2 = 1 + 4 + 1 + 4 + 0 + 1 = 11$
(d) $N = 6$
(e) $\bar{D} = 7/6 = 1.17$

$$S_{\bar{D}} = \frac{S_D}{\sqrt{N}} = \sqrt{\frac{N\Sigma D^2 - (\Sigma D)^2}{N^2(N-1)}}$$

Plugging in the components needed for $S_{\bar{D}}$, we have

$$S_{\bar{D}} = \sqrt{\frac{6(11) - 49}{36(5)}} = .31$$

Thus

$$t = \frac{1.17}{.31} = 3.81$$

Step 4.   3.81 > 2.571; therefore reject H_0.

The investigator concludes (assuming an adequate experimental design) that the short-term memory span is also adversely affected by this type of amnesia.

Assumptions Involved in Doing a *t* Test

The development of the *t* test is based on random samples being drawn from a normally distributed population. Therefore, in a technical sense, to have a valid *t* test we must assume that the populations representing our experimental conditions are (a) normally distributed and (b) have equal variances. However, the *t* test will give reliable results with moderate violations of these assumptions, particularly with large N. A further discussion of this matter is presented in Chapter 12.

A NOTE ON RETAINING H_0

Based on the statistical model that we have been using, if we use $\alpha = .05$ and reject H_0 based on a statistical test, we can have 95% (i.e., $1 - \alpha$) confidence in our decision. However, if we retain H_0, we cannot say with a certain confidence that H_0 is really true. All we can say is that there is not sufficient evidence to reject H_0. The two are not the same. An example from real life may clarify this seeming paradox. Suppose you rent a large old house by the beach for the summer. Although a generous person, you do not like sharing domestic quarters with one of earth's oldest inhabitants, the cockroach. Therefore, you set up H_0 as there are *no* roaches in the house. If you find evidence that H_0 is false, you will spend the necessary money for an exterminator. If you *retain* H_0, you will keep your money and assume there are no roaches in the house. Next, you test H_0 by examining in, under, and around likely places that roaches may hide. After a careful search, you find not a single roach, so you retain H_0. Notice, however, that when you retained H_0, you did not prove it to be true.

There may be roaches in places that you could not get to. In effect, what you did when you retained H_0 was to say, ''I will act as if H_0 were true until I find evidence to the contrary.'' The same analogy is true for the researcher. When the researcher retains H_0 from a statistical test, the researcher acts as if H_0 were really true while realizing that further experimentation or using a more sensitive experimental design might well show the effect of the experimental manipulation that he or she failed to find.

THE COMPUTED _t_ VALUE AND EXPERIMENTAL SENSITIVITY

It should now be apparent that the larger the computed _t_ value, the greater the probability of rejecting H_0. If an independent variable is really effective, an investigator wants to be able to show that this is so. The objective is then to let the computed _t_ value become as large as possible. (There are some constraints on this philosophy, which we will discuss in Chapter 12.) Factors that make the computed _t_ value large are the same factors that make any experimental design more sensitive. By _sensitive_ we mean the ability to detect accurately small differences between sets of data. How can this be done? Let us look at the three components that affect the size of a computed _t_ value. Basically, _t_ is determined by the difference between means divided by a term reflecting the variability of the scores within conditions of the experiment. The latter is divided by a term reflecting sample size. These relationships can be conceptualized by using Equation 6.4:

$$t = \frac{\bar{X}_1 - \bar{X}_2}{\sqrt{\dfrac{S_1^2}{N_1} + \dfrac{S_2^2}{N_2}}}$$

Thus, _t_ will increase if the

(a)   difference between means increases
(b)   sample size increases
(c)   variability within the sets of scores representing conditions decreases

Of these three variables that affect the size of a computed _t_ value, an experimenter has complete control over sample size and partial control over the variability of scores within conditions. The experimenter (provided he or she is honest) has control over the magnitude of the difference between means only in situations where the independent variable can be represented by conditions that can vary in quantity. Drug dosage would represent such a situation. Different therapy techniques would be represented by conditions that are different, but not by a quantifiable amount. To maximize the chances of getting a large, and therefore significant, computed _t_ value, an experimenter should avoid small sample sizes and design the experiment such that the variability within sets of scores will be as small as possible. The latter can be accomplished by, for example, (a) giving clear consistent instructions to all subjects (when subjects are humans), (b) having reliable equipment, (c) having an environment that is adequate for the task and free of distractions, (d) striving to obtain a constant and high motivation level for all subjects.

The importance of these principles is shown by examining what might happen if one of them is seriously violated. Let us use our drug-versus-placebo reaction time experiment in this example. Suppose we try to do this experiment in a noisy environment where people are slamming doors, yelling in the hall, and so on. If the environment happened to be quiet just as the subject was responding to the reaction time stimulus onset, he or she might get a fast time, but if a loud noise occurred just at that moment it might slow the reaction time considerably. Now if the noise and distractions occur randomly with both the drug and placebo groups, the mean score for each group will be raised about the same amount, so the difference between the means will remain about the same as if no distractions had occurred. However, a lot of additional variability (extra slow times) will occur within the sets of scores. The effect of this will be to make the Variance term large, thus reducing the computed t value and thereby reducing the probability of achieving statistical significance with the resulting t test. Therefore an experimenter must always try to eliminate or minimize the effects of extraneous variables.

CHOOSING DEPENDENT VERSUS INDEPENDENT GROUPS t TESTS

There is yet another way that the variance term in a t test can be reduced. It comes from the fact that people (or other organisms) are usually different with respect to the dependent variable. In our reaction time example, some people have, as a part of their physical makeup, a much faster reaction time potential than other people (for example, Muhammad Ali in his prime reportedly could deliver a left jab in 200 milliseconds). If we randomly assign people to groups, differences between people become an inherent part of the variability of the scores within each condition. In other words, fast (low scores) and slow (high scores) will occur in each condition irrespective of the effect of the drug. However, we can eliminate this source of variability by matching pairs of subjects with respect to the dependent variable or by having the same subjects take both conditions of the experiment. Let us use the latter case to illustrate. With the dependent groups t test, the data (in hundredths of a second) from which the computed t value is derived comes from computing the difference between conditions for the same subjects. In other words, how did each person do compared to himself or herself? Whether a person was slow or fast to start with is not picked up in the difference score (D) data. The data in Table 6.2 illustrate this.

The drug is consistently slowing everyone down about the same amount (i.e., from 4 to 6 hundredths of a second, so there is little variability within the D scores and, thus, a very small Variance term

TABLE 6.2
Illustrating the Advantage of a Repeated-Measures Design

	Placebo (X_1)	Drug (X_2)	D
S_1	31	36	+5
S_2	24	28	+4
S_3	29	35	+6
S_4	25	31	+6
S_5	33	37	+4
	$\bar{X}_1 = 28.4$	$\bar{X}_2 = 33.4$	$\bar{D} = 5$

The data in columns 1 and 2 show high variability, whereas the difference scores show low variability.

in the computed *t* value; thus increasing the size of *t*. However, if you look at the scores within each condition, you will see larger differences, reflecting inherent differences in people's reaction times. Thus, the variability component of the computed *t* value will be much greater for the independent groups *t* test than for the dependent groups *t* test. This would result in a much higher computed *t* value for the dependent groups *t* test. In fact, the actual computed *t* value using the numbers in this example for the independent groups *t* is 2.07, whereas the computed *t* value for the dependent groups *t* test is 11.18. A by-product of this point is that if you make a mistake and do the wrong *t* test, a serious error occurs. Only if the variability between subjects is small would the dependent groups and independent groups *t* test values from the same data be similar in size (although one test would still be incorrect). Ordinarily, in real experimental situations, the variability between subjects is large enough to create a considerably more sensitive experimental design by using repeated measures instead of independent groups.

The reason why dependent groups designs in the form of repeated measures on subjects are not always preferred over independent groups designs is that sometimes the participation in one condition of an experiment will influence the performance in another condition. For example, suppose the dependent variable is heart rate and the independent variable is opening a package with or without a dead snake in it. Opening a package after the snake experience would not be the same (even with nothing aversive inside) as before the snake experience. In other words, once initiated the fright is likely to carry over to subsequent conditions even if these conditions are not supposed to be frightening. *Carryover effects* are the major reason why independent groups designs are sometimes chosen instead of repeated-measures designs.

carryover effects:

the effects of having participated in one condition carry over into another condition (carryover effects may be a problem in repeated measures designs)

One point involving experimenter's decisions that we have discussed with respect to the *t* test should be remembered. All decisions that affect the size of a computed *t* value in a *t* test are valid for any statistical test of inference and will affect that test in the same way that they affected the *t* test (i.e., increasing the sample size will increase the size of the computed test statistic, and increased variability will reduce the size of the computed test statistic).

SUMMARY OF THE PARAMETRIC STATISTICAL INFERENCE MODEL

Now that we have developed the procedure for making inferences using the *t* test, let us review the model in a slightly more formal way. Table 6.3 presents a six-step summary of the statistical inference model. Notice that the population from which we take random samples does not have to be specified by having each member identified. Indeed, in most experiments the population referred to by the dependent variable is defined by a characteristic of an organism such as a measure reflecting learning, motivation, reaction time, anxiety level, and so on. Restrictions are usually placed on the organism studied (e.g., college students, males or females, 5-year-olds). The population of scores or potential scores in an experiment is usually, then, a measurable characteristic of some subpopulation of an organism.

Notice also that the inferential statistics model only allows us to say that a set of results likely did not happen because of random influences (if the null hypothesis is rejected). Attributing the results to the independent variable is an inference we make from our experimental design. See Box 3.1.

TABLE 6.3
The Statistics Model Using Independent Groups

Step A	Population representing dependent variable. Example: reaction time scores for young adults.
Step B	Random selection of elements that match the definition of A into groups that will represent conditions of the independent variable(s).
	Critical assumption from H_0: The only difference at this point in potential means (of size N) are likely small ones due to random influences.
Step C	Introduce levels of I.V.'s. If I.V. has no effect, the means will likely be similar.

TABLE 6.3 (Continued)

Step D	Do statistical test (e.g., *t* test). The result of the statistical test is a probability value that the obtained statistic (e.g., *t*) occurred with H_0 assumed to be true.
Step E	Decision: 1. If step D probability is small (less than .05), then reject H_0. This means that the difference between means was larger than would be expected through random samples drawn from the same population. 2. If step D probability is greater than .05, retain H_0.
Step F	Inference: 1. If H_0 is retained, we usually infer that our I.V. manipulation was ineffective—that is, the D.V. was not affected by it. 2. If H_0 is rejected, we can say that something other than chance created the differences between means. In a well-designed experiment, we infer this something to be the effects of the I.V.; for example, smoking marijuana slows down reaction time.

SUMMARY

The *t* distribution is a theoretical distribution, similar in shape to the standard normal distribution, that is an appropriate statistical model to use when population variances are unknown and have to be estimated from samples. The *t* test uses the *t* distribution to provide a cutoff value, which is used to determine whether to reject a sample mean (or difference between means) as a chance occurrence (i.e., whether to reject H_0). If the computed *t* value is larger than the cutoff *t* value, this implies that the difference between means arising from two sets of data likely did not happen because of random influences (i.e., the null hypothesis can be rejected). The critical *t* value is determined by three factors: (a) the degrees of freedom in the data, (b) the chosen α level, and (c) the directionality of the test. The formula for the computed *t* value for a dependent groups *t* test is

$$t = \frac{\bar{D}}{S_{\bar{D}}}$$

A dependent groups design occurs when (a) the same subjects participate in all of the conditions (two for the *t* test) of an experiment or (b) each subject in a condition has a matching subject in the other conditions of the experiment. An independent groups design occurs when different subjects participate in various experimental conditions. The independent groups t test differs from the dependent groups t test in two ways: (a) the critical t will be slightly smaller for the independent groups t test due to the larger number of degrees of freedom; (b) the denominator of the computed t for the two tests will be different. The independent groups t test reflects the variability within each set (e.g., condition) of scores rather than the variability within the difference scores that result when a score in one condition is subtracted from a dependent score in a second condition. The formula for the independent groups t test is

$$t = \frac{\overline{X}_1 - \overline{X}_2}{S_{\overline{x}_1 - \overline{x}_2}}$$

The denominator, $S_{\overline{x}_1 - \overline{x}_2}$ is called *the standard error of the difference between means for independent groups* and can be computed in two ways:

(a)   When $N_1 = N_2$,

$$S_{\overline{x}_1 - \overline{x}_2} = \sqrt{\frac{S_1^2}{N_1} + \frac{S_2^2}{N_2}}$$

(b)   When $N_1 \neq N_2$,

$$S_{\overline{x}_1 - \overline{x}_2} = \sqrt{S_{\text{pooled}}^2 \left[\frac{1}{N_1} + \frac{1}{N_2} \right]}$$

When $N_1 = N_2$, both methods give the same answer, but the first method is easier.

The first decision to make in doing any t test is to determine whether the dependent groups version or the independent groups version is appropriate. Failure to make the appropriate choice is a very serious error.

When we use a particular α level, we can have $1 - \alpha$ confidence in our decision when H_0 is rejected. However, we cannot say from a statistical test anything about the certainty of H_0 being true. As in our roach example, accepting that there are no roaches in the house is not proving that there are no roaches in the house.

There are two independent ways that a researcher can design an experiment to increase the sensitivity of the experiment and to increase the chance of achieving statistical significance: (a) increase the size of N in the study, (b) attempt to decrease the variability within the data. Two ways of accomplishing the latter are: (1) stabilize extraneous factors that might affect the data, (2) use a type of dependent groups design.

Dependent Groups Designs

Definitional Formula	Computational Formula

$$t = \frac{\bar{X}_1 - \bar{X}_2}{S_{\bar{D}}}$$

$$t = \frac{\bar{X}_1 - \bar{X}_2}{\sqrt{\dfrac{S^2}{N}}}$$

where $\dfrac{S^2}{N} = \dfrac{N\Sigma D^2 - (\Sigma D)^2}{N(N-1)N}$

N = No. of difference scores

Independent Groups Designs ($N_1 = N_2$)

Definitional Formula	Computational Formula

$$t = \frac{\bar{X}_1 - \bar{X}_2}{S_{\bar{x}_1 - \bar{x}_2}}$$

$$t = \frac{\bar{X}_1 - \bar{X}_2}{\sqrt{\dfrac{S_1^2}{N_1} + \dfrac{S_2^2}{N_2}}}$$

where $S^2 = \dfrac{N\Sigma X^2 - (\Sigma X)^2}{N(N-1)}$

N = No. of scores in a group

Independent Groups Designs $N_1 \neq N_2$

Definitional Formula	Computational Formula

$$t = \frac{\bar{X}_1 - \bar{X}_2}{S_{\bar{x}_1 - \bar{x}_2}}$$

$$t = \frac{\bar{X}_1 - \bar{X}_2}{\sqrt{\dfrac{SS_1 + SS_2}{N_1 + N_2 - 2}\left[\dfrac{1}{N_1} + \dfrac{1}{N_2}\right]}}$$

where $SS = \dfrac{N\Sigma X^2 - (\Sigma X)^2}{N}$

N = No. of scores in a group

Note. The unequal N formula can also be used (and will give the same answer) for equal N situations

Statistical Calculators

With calculators that are programmed to compute standard deviations, the following procedures are time savers.

a. S^2 = the square of the standard deviation (i.e., S).
b. $SS = (N - 1)S^2$

Values for S^2 or SS can then be plugged into the appropriate computational formula.

KEY TERMS

theoretical sampling distribution

standard normal distribution

t distribution

critical t

computed t

one-tailed test

two-tailed test

dependent groups

repeated measures

independent groups t test

dependent groups t test

S_D

$S_{\bar{x}_1 - \bar{x}_2}$

simulation

pooled variance

conservative statistical test

carryover effects

sensitivity

PROBLEMS

1. Estimate whether the following data would show significantly different (i.e., reject H_0) means. Then do a dependent groups t test, using $\alpha = .05$ for a nondirectional test.

	Group 1	Group 2
S_1	17	21
S_2	15	18
S_3	21	22
S_4	13	18
S_5	14	16

2. Suppose the data in problem 1 came from 10 people, 5 in each group. Guess at a t value and whether it is significant; then do an independent groups t test on the data, using $\alpha = .05$ for a nondirectional test.

3. Verify the accuracy of the computed t values given for the dependent groups and independent groups t tests that were discussed in the section on experimental sensitivity. Then, using $\alpha = .05$, nondirectional test, determine for both computed t values whether H_0 can be rejected.

4. Using four numbers in each group, make up a set of data that would be nonsignificant with an independent groups t test, but, using that same set of data, significant with a dependent groups t test. Do the t tests to verify. Use $\alpha = .05$ for a nondirectional test. [*Hint:* Read the section on experimental sensitivity, pp. 131–132.]

5. Using Table A, the random number table, pick three numbers at random (e.g., the first three numbers in a column) and find the mean

of these scores. Pick a different random group of three numbers (e.g., the next three numbers from the same column as before) and find the mean of these three numbers. Now subtract the mean of the first set from the mean of the second set and record the difference. Repeat this procedure 15 times, using different sets of three random numbers.

 a   Find S for these 15 difference between means (where $N = 3$) scores. This value is $S_{\bar{X}_1 - \bar{X}_2}$ for these scores.

 b   In experiments, $S_{\bar{X}_1 - \bar{X}_2}$ is usually found by using Equation 6.3. Use this formula to find $S_{\bar{X}_1 - \bar{X}_2}$, using the two sets of three random numbers that you first chose. The value of $S_{\bar{X}_1 - \bar{X}_2}$ computed in this manner will likely be close but not the same as the value for $S_{\bar{X}_1 - \bar{X}_2}$ in (a). Why?

 c   Round these 15 scores from (a) to the nearest whole number. Then make a histogram (Chapter 2) of these scores. Around what value does your histogram appear to be centered? Why?

6. Make up a hypothesis that implies a directional test and another that implies a nondirectional test.

7. Do independent groups t tests on the following. Use $\alpha = .05$ for a nondirectional test.

 a $\bar{X}_1 = 15$, $\bar{X}_2 = 19$, $S_1^2 = 10$, $S_2^2 = 13$; $N_1 = 6$, $N_2 = 6$
 b $\bar{X}_1 = 15$, $\bar{X}_2 = 19$, $S_1^2 = 10$, $S_2^2 = 13$; $N_1 = 36$, $N_2 = 36$
 c $\bar{X}_1 = 15$, $\bar{X}_2 = 19$, $S_1^2 = 100$, $S_2^2 = 169$; $N_1 = 36$, $N_2 = 36$
 d $\bar{X}_1 = 15$, $\bar{X}_2 = 25$, $S_1^2 = 100$, $S_2^2 = 169$; $N_1 = 36$, $N_2 = 36$

 e   What happens to the computed t value (a) as N_1 and N_2 get larger? (b) as S_1^2 and S_2^2 get larger? (c) as $\bar{X}_1 - \bar{X}_2$ gets larger?

8. Describe an experimental situation (not in the chapter) where (a) a dependent groups t test would be preferred; (b) an independent groups t test would be preferred. State your reasons.

9. Make up one set of data such that a sign test on these data would be nonsignificant but a dependent groups t test on the same data would be significant. Do each test to verify. Use $\alpha = .05$ for a nondirectional test.

10. An investigator wished to find out if any difference existed between the spelling capabilities of children taught either by the rote repetition method or by the phonic method. Children were randomly assigned to groups. After 10 weeks of instruction, a spelling test was given and the following scores obtained. Do the appropriate (dependent or independent?) t test, using $\alpha = .05$.

Rote scores: 79, 81, 39, 65, 89, 75, 73, 91, 65, 77
Phonic scores: 83, 95, 65, 79, 94, 88, 85, 93, 77, 90

11. An investigator in a sleep laboratory wants to know if eating a large meal before going to bed will affect the number of dreams that a person has during 8 hours of sleep. Eight subjects agree to participate in the study. Each subject participates twice, once after eating a large meal before going to bed (condition E), and once after eating nothing for 5 hours before going to bed (condition N). The two sessions are given one week apart to minimize carryover effects. Four of the subjects eat before the first session, and four do not eat before the first session to equalize habituation effects. The data in terms of numbers of dreams were as follows. Do the appropriate t test, using $\alpha = .05$ for a nondirectional test.

	Condition	
	E	N
S_1	7	3
S_2	5	4
S_3	6	6
S_4	8	6
S_5	4	5
S_6	7	7
S_7	9	5
S_8	7	4

12. An investigator wished to test the effects of feedback on a strength task. In one condition (F), subjects could see a dial indicating the strengths in pounds of their grip in their preferred hand. In the other condition (N), subjects could not see the dial. Twenty male subjects were randomly assigned into the two conditions. Test the following data for significance, using $\alpha = .05$ for a nondirectional test.

F: 180, 150, 95, 125, 188, 210, 173, 168, 171, 147
N: 195, 143, 158, 171, 138, 88, 145, 191, 130, 141

13. An investigator wished to see if boys would be differently affected by competition than girls. Twelve fifth-grade boys were matched in terms of spelling ability with 12 fifth-grade girls. Each matched pair had to compete against each other in a boys-versus-girls spelling contest. The number of words correctly spelled by boys and girls by pair is given below. Is there a significant difference between boys and girls? Use $\alpha = .05$ for a nondirectional test.

Pair	Boys	Girls
1	10	10
2	9	8
3	9	7
4	5	8
5	9	7
6	8	10
7	6	9
8	7	7
9	5	8
10	1	6
11	3	7
12	1	4

14. An investigator predicts that a biofeedback group will be able to lower its blood pressure more during a 30-minute session than will a control group that is given the same treatment, except that it does not see a blood pressure dial. Thirty subjects were randomly assigned into the two groups. Based on the following change score data for systolic blood pressure, was the investigator's hypothesis supported? Use α = .05 for a nondirectional test. Report the results of the t test with a one-sentence description as you might see it in a journal article.

Change Scores (beginning score minus end score)

Feedback group: 8, 7, 3, 7, 1, 6, 10, 17, −4, 11, 0, 8, 5, 3, 9
Control group: 6, 6, −3, 0, 5, 0, 8, 12, 2, 7, 7, 1, 3, 4, 1

15. An investigator predicts that a temporary "repression effect" will cause the recall of emotionally charged words to be greater after a 20-minute delay than immediately following the list presentation. Twelve subjects were presented with a list of 30 emotional (as determined by independent ratings) words. The words were shown one at a time on a screen at a 5-second rate. Following the presentation of the list, subjects were asked to write on a sheet of paper all of the words that they could remember. After solving puzzles for 20 minutes, they were again tested for recall of the list. Using a nondirectional t test with α = .05, determine if the investigator's prediction was supported. (Be careful, this is tricky.) Report the results of the test as you would find it in a journal article.

	1st test	2nd test
S_1	15	12
S_2	21	19
S_3	17	11
S_4	18	19
S_5	25	25
S_6	21	14
S_7	20	15
S_8	17	12
S_9	14	10
S_{10}	19	16
S_{11}	22	16
S_{12}	24	15

16. An investigator hypothesizes that first-grade boys will show more aggressive actions in a free-play situation than will first-grade girls. Observers record the number of aggressive acts during a 20-minute recess. Use $\alpha = .05$ for a nondirectional test to test the researcher's hypothesis. State whether the researcher's hypothesis was supported and then report the results of the test in journal format.

Aggressive Acts

Boys: 7, 3, 1, 0, 0, 7, 0, 2, 1, 0, 4, 0, 1, 3
Girls: 0, 1, 8, 0, 0, 2, 0, 4, 0, 1, 1, 0

CHAPTER 7

One-Factor Analysis of Variance

The *t* test is an appropriate way to test for significant differences between the means of two conditions. The shortcoming of the *t* test is that it can only be used for a test between two conditions. If you want to do a single test comparing the means of three or more conditions, then you must use a different test. We will now discuss such a test, developed by R. A. Fisher, called the *analysis of variance* (ANOVA). The ANOVA can be utilized with more than one independent variable (i.e., factor), but our concern in this chapter will be limited to the one-factor situation. Remember that all statistical tests of inference use the same principles in a similar fashion. Therefore, let us briefly review the development of the t test, and then we will examine the new test (ANOVA), which utilizes a family of theoretical distributions called the F distributions.

REVIEW OF THE t TEST

The basic idea of all statistical tests of inference is to find out what sorts of outcomes one might expect under conditions where the null hypothesis (H_0) is true. Then, if the outcome from a particular test deviates enough from the expected outcome when H_0 is true, H_0 is rejected and we infer from our experimental design that the independent variable was creating the differences that we see between sets of data. The t test for independent groups and $N_1 = N_2$ took the form

$$t = \frac{\bar{X}_1 - \bar{X}_2}{\sqrt{\dfrac{S_1^2}{N_1} + \dfrac{S_2^2}{N_2}}}$$

If H_0 were true, we would expect t values within a certain range. This range would depend specifically on the size of N, the α level, and the directionality of the test, but generally one would expect t values between $+2$ and -2. If our computed t value were outside of this range (i.e., larger than the critical t value), we would reject H_0. We were able to determine the critical t value because we knew that the theoretical frequency distribution called the t distribution (for a given number of degrees of freedom) was a close approximation of the frequency distribution that would result after thousands of replications where t scores were computed under conditions where H_0 was true. In other words, the t distributions provided a theoretical statistical model for us to use and with which to make decisions about whether to reject H_0 given two sets (i.e., conditions) of empirical data.

What we need now is a different theoretical frequency distribution and a different statistical test. We need a test that will be appropriate for testing differences between the means of several conditions, not just two. Let us develop such a test by looking at the components of the t test in a different way.

THE ONE-FACTOR ANALYSIS OF VARIANCE

Suppose we look at the numerator in the t test, $\bar{X}_1 - \bar{X}_2$, and notice that the bigger the difference between the means, the more variability there is between the two means. Remember, variability means score spread. The farther apart two scores are, the more score spread there is; hence, the greater is the variability. Now suppose we express this score spread between means in terms of S^2 (see Chapter 3). Remember,

$$S^2 = \frac{SS}{N-1} = \frac{\Sigma (X - \bar{X})^2}{N-1}$$

so we compute S^2 by using the two means as the two X scores. If we take the S^2 of the two means and multiply this by N (the number of scores from which each mean was derived) we have a term called *means squares between* (MS_B) and it will be the numerator of our new test. The denominator will be the *sum of squares* (SS) within each set of scores added together and divided by the sum of the degrees of freedom within each group of scores. This term is called *mean squares within* (MS_W), and it reflects variability within sets of scores. Box 7.1 illustrates relative values of MS_B and MS_W.

BOX 7.1
Conceptual Illustration of MS_B and MS_W

	Panel a		Panel b	
	A	B	C	D
	12	21	15	6
Variability within	10	20	5	18
conditions is	8	20	0	12
reflected in MS_W	9	18	20	20
	11	21	10	4

Variability within conditions is reflected in MS_W

$\bar{X}_A = 10$ $\bar{X}_B = 20$ $\bar{X}_C = 10$ $\bar{X}_D = 12$

Variability between means of conditions is reflected in MS_B

Variability between means of conditions is reflected in MS_B

Observe that MS_B will be larger in panel a compared with panel b, and MS_W will be larger in panel b compared with panel a.

The ratio of MS_B/MS_W is called an **F ratio.** Hence,

$$F = \frac{MS_B}{MS_W}$$

Notice that the characteristics (i.e., larger N and larger difference between means) that make the computed value for the F test larger also had the same effect on the t test. Also, a larger variance within conditions (i.e., S_1^2 and S_2^2) made t smaller, as it makes the computed F value for the new test smaller. In fact, for two conditions (two groups or sets of data), the relationship between a t value and an F value is a precise one given by the formula $t^2 = F$. If one computed a t

value, then squared it, one will get the same value as MS_B/MS_W, which equals F, for that data.

The next question is, "What kind of values should one get from the ratio of MS_B/MS_W if H_0 is true?" We could answer this question as we did with the t test by doing a simulation where each condition in our design is the same (e.g., a placebo), thus ensuring that H_0 is true. For example, in a study of the effects of a drug on reaction time, all groups might be given placebos. We then compute MS_B/MS_W, obtain a number, and repeat the procedure several hundred times. The numbers representing MS_B/MS_W would form a frequency distribution from which we could determine the rare (occur less than .05 of the time) scores. If we then have data from a real experiment where the value for MS_B/MS_W was larger than the cutoff representing the boundary for a rare event (this cutoff being obtained from our simulation where H_0 was true), we would reject H_0. The theoretical distribution that would match the shape of this simulated distribution is one of the F distributions, and we use it, as we used the t distribution, to find a critical (i.e., cutoff) value from which we can make a decision about rejecting H_0. We therefore compute

$$F = \frac{MS_B}{MS_W}$$

We then compare our computed F value with the critical F value and make our decision about rejecting H_0. We will study the F distributions and critical F's in more detail later. For now we note the advantage of the F test that has been developed using ANOVA. We can now compute an F value for testing H_0 between any number of conditions, since both the numerator (MS_B) and denominator (MS_W) have no size constraints in terms of the number of conditions that can be analyzed. Before we turn our attention to the theoretical F distributions and finding critical F values, let us review the development of MS_B and MS_W in a more formal way that can be expanded into tests involving more than one independent variable.

PARTITIONING SUMS OF SQUARES WITH EQUAL *N* PER CONDITION

We will develop procedures for finding MS_B and MS_W within the context of an experiment. Suppose we are interested in the effects of amounts of caffeine on a motor skills task. We set up three levels of caffeine: none, medium, and high represented by giving subjects a placebo, one, or three caffeine pills. We measure performance by the number of errors committed while performing the task. We have four different randomly chosen subjects in each condition. (In a real experiment, we would want 20 or more subjects per condition. We use four for ease of illustrating calculations.) The data are given in Table 7.1.

TABLE 7.1

Placebo	Medium (1 pill)	High (3 pills)
3	6	12
2	2	8
4	3	6
3	5	6
$\bar{X} = 3$	$\bar{X} = 4$	$\bar{X} = 8$

$\bar{X}_G$ (mean of all 12 scores) $= 5$

The subscript G stands for grand mean, which is interpreted as the mean of all scores in all conditions. We have four scores in each of three conditions for a total of 12 scores. If we consider the 12 scores as *one group* of scores and compute the sum of squares (SS) using all 12 scores, we have the *total sum of squares* (SS_T).

Recall again the following from Chapter 3:

$$SS = \Sigma (X - \bar{X})^2$$

also:

$$S^2 = \frac{\Sigma(X - \bar{X})^2}{N - 1} \quad \text{or} \quad \frac{SS}{N - 1}$$

Therefore, the formula for SS_T says to take each score within each condition, subtract the grand mean from it, square the resultant scores, and add them. In our example,

$$SS_T = (3-5)^2 + (2-5)^2 + (4-5)^2 + (3-5)^2$$
$$+ (6-5)^2 + (2-5)^2 + (3-5)^2 + (5-5)^2$$
$$+ (12-5)^2 + (8-5)^2 + (6-5)^2 + (6-5)^2 = 92$$

It can be shown that SS_T for any number of groups of scores can be broken down into two independent components such that

$$SS_T = SS_B + SS_W$$

where

$$SS_B = N_c \Sigma(\bar{X}_c - \bar{X}_G)^2$$

and

$$SS_W = \Sigma(X - \bar{X}_c)^2$$

Note: $\bar{X}_c$ represents the mean of a condition; N_c represents the **number of scores within a condition.**

The verbal interpretation of SS_B is, "Take the mean of each condition, subtract the grand mean ($\overline{X}_G$) from it, square each resultant value, and add these values. Then multiply that total by the number of scores in a condition (i.e., N_c)." The formula for SS_W says, "Take each score and subtract the mean of that score's condition (i.e., $\overline{X}_c$) from that score. Then square the scores resulting from the subtraction (deviation scores) and add them."

Notice that SS_W is each score minus the mean of the *condition* containing that score, squared and added for all scores. This is simply $SS_1 + SS_2 + \cdots + SS_k$, where the subscripts designate the conditions, with k being the last condition. To illustrate for our example:

$$SS_B = 4[(3-5)^2 + (4-5)^2 + (8-5)^2] = 56$$

$$SS_W = (3-3)^2 + (2-3)^2 + (4-3)^2 + (3-3)^2$$
$$+ (6-4)^2 + (2-4)^2 + (3-4)^2 + (5--4)^2$$
$$+ (12-8)^2 + (8-8)^2 + (6-8)^2 + (6-8)^2$$
$$= 36$$

These values for SS_B (56) and SS_W (36), add to the value obtained for SS_T (92). We next need to find the degrees of freedom appropriate for SS_B and SS_W.

The degrees of freedom can be broken down just as the sum of squares were broken down:

$$df_T = df_B + df_W$$

where $df_T = N_T - 1$ (i.e., numbers of scores in all conditions minus 1)

$df_B = k - 1$ (i.e., number of conditions minus 1)

$df_W = \Sigma(N_c - 1)$ (i.e., the number of scores in a condition minus 1 summed across conditions).

In our example, for equal N, the sum sign can be replaced by N_c.

$$df_T = 12 - 1 = 11$$

$$df_B = 3 - 1 = 2$$

$$df_W = (4 - 1) + (4 - 1) + (4 - 1) = 9$$

The relationship between the various mean squares (MS), sum of squares (SS), and degrees of freedom is

$$MS_B = \frac{SS_B}{df_B} \quad \text{and} \quad MS_W = \frac{SS_W}{df_W}$$

Since $F = MS_B/MS_W$, we first find MS_B and divide that by MS_W. In our example,

$$MS_B = \frac{56}{2} = 28 \quad \text{and} \quad MS_W = \frac{36}{9} = 4.0$$

Therefore,

$$F = \frac{28}{4.0} = 7.0$$

EXPECTED *F* VALUE WHEN H_0 IS TRUE

There is a rationale for why the ratio of MS_B/MS_W is an appropriate one to test H_0. It can be demonstrated that given conditions (a) where the null hypothesis is really true (such as giving all groups placebos), and (b) where the population from which random samples are taken (representing data within conditions) is normally distributed, MS_B and MS_W will yield the same expected value (i.e., same mean value with infinite replications). However, because of random variations in sampling (i.e., sampling error), any sample of scores representing conditions will contain random differences between MS_B and MS_W. In other words, when H_0 is true, a sampling distribution will arise from MS_B/MS_W such that the most frequent occurrence will be when MS_B and MS_W have approximately the same value. Since any number divided by itself equals 1, the most frequent values of MS_B/MS_W will be close to 1. However, if H_0 is false, then the differences between the means of the conditions are reflected in MS_B but not MS_W. Thus, when H_0 is false, MS_B will be expected to be larger than MS_W and MS_B/MS_W will be greater than 1. We can show these principles involving MS_B and MS_W algebraically as follows:

$$MS_B = \text{error} + \text{treatment effects}$$

$$MS_W = \text{error}$$

$$F = \frac{MS_B}{MS_W} = \frac{\text{error} + \text{treatment effects}}{\text{error}}$$

treatment effects:

effects in an experiment due to an independent variable

error term:

the denominator in an *F* test; it reflects the variability within cells

where error means random variation within conditions, and *treatment effects* refer to the effects from the independent variable. Because the denominator of the *F* ratio contains only ''error'' (i.e., random variation within conditions), the term that reflects the denominator (e.g., MS_W) is called the *error term.* Thus, to repeat, if there are no treatment effects, the expected *F* value will be 1. If treatment effects exist, *F* will be greater than 1, the magnitude of *F* depending on the size of the treatment effects.

For our sample experiment, we have computed an *F* value of 7.0. We next need to know the appropriate sampling distribution for *F* under conditions where H_0 is true. We will then mark off 5% (for $\alpha = .05$) of the area in the tail that represents unlikely F values to occur when H_0 is true. These values have been found for us by mathematicians, and the critical F's (i.e., the F's that divide the area into a 95%-5% split) are listed in Table D. Thus, if our computed F value is larger than the critical F, we will reject H_0.

THE *F* DISTRIBUTIONS

The F distributions are a family of mathematical curves having two parameters. With respect to using an F distribution to find a critical F in a statistical test, one parameter (listed across the top of an F table) is df_B. The second parameter (listed down the left side of an F table) is df_W. The F distributions have the following properties:

(a) F values range from zero to positive infinity.
(b) F distributions are not symmetrical, but skewed to the right.
(c)   The F distributions become more normal in shape as the values of both parameters become larger.
(d)   The F distributions get increasingly closer to the x axis with values greater than 1.

The particular F distribution that we wish to use as a statistical model from which to make a decision about rejecting H_0 in our current experiment is the distribution having 2 df between groups (df_B) and 9 df within groups (df_W). This distribution is shown in Figure 7.1, with 5% of the area in the tail marked off.
A computed F value that falls within the rejection region represented by an α level is cause for rejecting H_0.

Figure 7.1
F distribution for 2 df and 9 df.

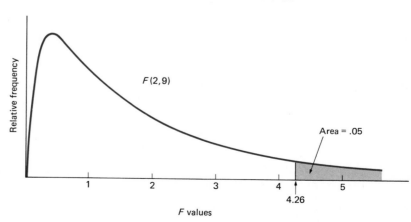

DIRECTIONALITY AND THE *F* TEST

Recall that a *t* value could be plus or minus, depending on the sign of $\bar{X}_1 - \bar{X}_2$. Thus a *t* value for a nondirectional test that is far enough in either tail would be reason to reject H_0. However, the *F* test is different. With the *F* test, any effects of the independent variable will show up as squared values in MS_B. Thus, with respect to MS_B, whether differences between means are positive or negative will make no difference. In either case a positive MS_B will result (given that the independent variable is effective), with an expected *F* value greater than 1. Therefore, the *F* test is always nondirectional with respect to the *F* distribution. Also, the critical *F* is always determined based on all of the α rejection area being in the skewed side of the distribution. All of the area to the left of the critical value, including values close to zero, is within the region where H_0 is retained. Therefore, if one wishes to use the *F* distribution to test a directional hypothesis (e.g., $\bar{X}_1 \geqslant \bar{X}_2$), a critical *F* (using $\alpha = .05$) must be chosen that divides the *F* distribution into a 90%-10% split. In other words, for a directional test, the critical *F* is found by doubling the α level that is stated in the layout of the statistical test. Since directional hypotheses are usually not made with more than two conditions (and rarely even with two conditions), we will use only nondirectional tests in conjunction with ANOVA.

PROCEDURE FOR DOING A ONE-FACTOR ANOVA

We will now outline a procedure for doing a one-factor ANOVA and then illustrate the outline with our experimental example. Note that H_0 is always defined as the equality of the means of all conditions (i.e., $H_0 : \mu_1 = \mu_2 = \mu_3 = \cdots = \mu_k$). We can now use a four-step outline as with the *t* test.

Step 1. Determine the α level and the degrees of freedom in MS_B and MS_W.

Step 2. From a table containing critical *F* values, find the appropriate critical *F* by finding the column represented by df_B and the row represented by df_W. Most tables, including Table D in the back of this book, will give at least two critical *F* values for each $df_B - df_W$ combination, representing levels of .05, .01 and perhaps others.

Step 3. Find the computed *F* value.

Step 4. If the computed *F* value is larger than the critical *F* value, reject H_0. If not, retain H_0, where H_0 is defined as "H_0 represents $\mu_1 = \mu_2 = \mu_3 = \cdots = \mu_k$."

Computing *F* Values

Since the principles involved in a one-factor ANOVA will generalize to more complex designs, you should learn the breakdown of the various SS values using the definitional formulas. This will give you the greatest understanding of the principles involved in ANOVA. However, as with the t test, these formulas are not the easiest to work with in doing the computations with most calculators. As before, the difference between the computational formulas and the definitional formulas is that an algebraic equivalent of $\Sigma(X - \bar{X})^2$ is used in computational formula.

We need to find values for SS_T, SS_B, and SS_W. Box 7.2 gives formulas for this. Additionally, Box 7.3 gives computational formulas for calculators that automatically compute a standard deviation. We will illustrate both the definitional formula and the computational formula in Box 7.2 by using them to find SS_T, SS_B, and SS_W.

Definitional Formula Example

Repeating the data from our caffeine experiment, we have

Placebo (X_1)			Medium (X_2)			High (X_3)		
X_1	$X_1 - \bar{X}_1$	$(X_1 - \bar{X}_1)^2$	X_2	$X_2 - \bar{X}_2$	$(X_2 - \bar{X}_2)^2$	X_3	$X_3 - \bar{X}_3$	$(X_3 - \bar{X}_3)^2$
3	0	0	6	+2	4	12	+4	16
2	−1	1	2	−2	4	8	0	0
4	+1	1	3	−1	1	6	−2	4
3	0	0	5	+1	1	6	−2	4

$$\Sigma X_1 = 12 \qquad\qquad \Sigma X_2 = 16 \qquad\qquad \Sigma X_3 = 32$$
$$\bar{X}_1 = 3 \qquad\qquad \bar{X}_2 = 4 \qquad\qquad \bar{X}_3 = 8$$
$$\Sigma(X_1 - \bar{X}_1)^2 = 2 \qquad \Sigma(X_2 - \bar{X}_2)^2 = 10 \qquad \Sigma(X_3 - \bar{X}_3)^2 = 24$$

Phase 1   with Steps

(a)   Sum scores in each condition (see preceding table).

(b)   Find the mean for each condition (see preceding table).

(c)   Find SS [i.e., $\Sigma(X - \bar{X})^2$] for each condition (see preceding table).

(d)   Count the scores in each condition: $N_1 = 4$; $N_2 = 4$; $N_3 = 4$.

(e)   Find the grand mean ($\bar{X}_G$) by finding the mean for all scores:

$$\bar{X}_G = \frac{3+2+4+3+6+2+3+5+12+8+6+6}{12} = 5$$

(f)   Find SS_T by finding SS, using all scores as a single group (i.e., $\Sigma(X - \bar{X}_G)^2$; that is, subtract the grand mean from each score, square the difference, then sum these squared differences).

$$SS_T = (3-5)^2 + (2-5)^2 + (4-5)^2 + (3-5)^2 + (6-5)^2 + (2-5)^2$$
$$+ (3-5)^2 + (5-5)^2 + (12-5)^2 + (8-5)^2 + (6-5)^2 + (6-5)^2 = 92$$

(g) Find SS_W by finding SS for each condition (i.e., the values in step c) and then summing the SS values [i.e., $\Sigma(X - \bar{X}_c)^2$].

$$SS_W = 2 + 10 + 24 = 36$$

(h) Find SS_B by finding SS using the means as a single group of scores, and then multiplying by the number of scores in a condition [i.e., $N_c \Sigma(\bar{X}_c - \bar{X}_G)^2$].

$$SS_B = 4[(3-5)^2 + (4-5)^2 + (8-5)^2] = 56$$

(i) Check:

$$SS_T = SS_B + SS_W$$

$$92 = 56 + 36$$

Phase 2 Find df_T, df_B, and df_W:

$$df_T = N_T - 1 = 11$$

$$df_B = k - 1 = 2 \text{ where } k \text{ is the number of conditions}$$

$$df_W = k(N_c - 1) = 3(4 - 1) = 9$$

$$df_T = df_B + df_W$$

Check: $11 = 2 + 9$

Phase 3 Compute *F*:

$$MS_B = \frac{SS_B}{df_B} = \frac{56}{2} = 28$$

$$MS_W = \frac{SS_W}{df_W} = \frac{36}{9} = 4.0$$

$$F = \frac{MS_B}{MS_W} = \frac{28}{4.0} = 7.0$$

Computational Formula Example

Repeating the data, we have

Placebo (X_1)	Medium (X_2)	High (X_3)
3	6	12
2	2	8
4	3	6
3	5	6

Phase 1

(a) Sum scores in each condition:

$$\Sigma X_1 = 12 \qquad \Sigma X_2 = 16 \qquad \Sigma X_3 = 32$$

Box 7.2
Formula for Computing Components of One-Factor ANOVA with Equal N and Unequal N per Condition

Equal N

Definitional Formulas

$$SS_T = \Sigma(X - \bar{X}_G)^2$$

That is, take each score, subtract the grand mean from it, square this difference, and add for all scores.

$$SS_B = N_c \Sigma[(\bar{X}_c - \bar{X}_G)^2]$$

That is, subtract the grand mean from the mean of each condition and square. Do this for all conditions and add; then multiply by the number of scores per condition.

$$SS_W = \Sigma(X - \bar{X}_c)^2$$

That is, from each score subtract the mean of the condition that that score is in, square the difference, and add for all scores.

Computational Formulas

$$SS_T = \frac{N_T \Sigma X^2 - (\Sigma X)^2}{N_T}$$

where X = each score in all conditions; N_T = total number of scores in all conditions.

$$SS_B = \frac{(\Sigma X_1)^2 + (\Sigma X_2)^2 + \cdots + (\Sigma X_k)^2}{N_c} - \frac{(\Sigma X_T)^2}{N_T}$$

where ΣX_1 = the sum of all scores in condition 1; ΣX_2 = the sum of all scores in condition 2, etc.; ΣX_T = sum of all scores in all conditions; k = the last condition; N_T = total number of scores in all conditions; N_c = number of scores in each condition.

$$SS_W = \Sigma X_T^2 - \frac{(\Sigma X_1)^2 + (\Sigma X_2)^2 + \cdots + (\Sigma X_k)^2}{N_c}$$

where ΣX_T^2 = sum of all squared scores in all conditions; the formula after the minus sign is identical to the first part of SS_B.

Unequal N

$$SS_T = \Sigma(X - \bar{X}_G)^2$$

That is, take each score, subtract the grand mean from it, square this difference, and add for all scores.

$$SS_B = \Sigma N_c (\bar{X}_c - \bar{X}_G)^2$$

For each condition, subtract the grand mean from the mean of that condition and square the difference. Multiply the result of this square by the number of scores in that condition. Sum this total for all conditions.

$$SS_W = \Sigma(X - \bar{X}_c)^2$$

That is, from each score subtract the mean of the condition that that score is in, square the difference, and add for all scores.

$$SS_T = \frac{N_T \Sigma X^2 - (\Sigma X)^2}{N_T}$$

where X = each score in all conditions; N_T = total number of scores in all conditions.

$$SS_B = \left[\frac{(\Sigma X_1)^2}{N_1} + \frac{(\Sigma X_2)^2}{N_2} + \cdots + \frac{(\Sigma X_k)^2}{N_k}\right] - \frac{(\Sigma X_T)^2}{N_T}$$

ΣX_1 = sum of all scores in condition 1;
ΣX_2 = the sum of all scores in condition 2, etc.;
k = the last group,
N_T = total number of scores in all conditions;
N_1 = number of scores in condition 1;
N_2 = number of scores in condition 2, etc.

$$SS_W = \Sigma X_T^2 - \left[\frac{(\Sigma X_1)^2}{N_1} + \frac{(\Sigma X_2)^2}{N_2} + \cdots + \frac{(\Sigma X_k)^2}{N_k}\right]$$

Note: The formula after the minus sign is identical to the first part of the formula for SS_B.

Box 7.2 (Continued)

Degrees of Freedom (Equal N)

$df_T = N_T - 1$ (i.e., total number of scores minus 1)

$df_B = k - 1$ (i.e., total number of conditions minus 1)

$df_W = k(N_c - 1)$ (i.e., the number of scores in a condition minus 1 times the number of conditions)

Degrees of Freedom (Unequal N)

$df_T = N_T - 1$ (i.e., total number of scores minus 1)

$df_B = k - 1$ (i.e., total number of conditions minus 1)

$df_W = \Sigma(N_c - 1)$ (i.e., for each condition subtract 1 from the number of scores in that condition and sum this result for all conditions)

Note: Unequal N formulas for SS and df are general and can be used for equal or unequal N. Equal N formulas are shortcuts appropriate only for equal N.

(b)   Square each sum for each condition:

$$(\Sigma X_1)^2 = 144 \qquad (\Sigma X_2)^2 = 256 \qquad (\Sigma X_3)^2 = 1024$$

(c)   Sum all scores in all conditions:

$$\Sigma X_T \quad (\text{all scores}) = 3+2+4+3+6+2+3+5+12+8+6+6 = 60$$

Check:

$$\Sigma X_T \ (\text{all scores}) = \Sigma X_1 + \Sigma X_2 + \Sigma X_3 = 12 + 16 + 32 = 60$$

(d)   Square the sum for all scores in all conditions:

$$(\Sigma X_T)^2 = 3600$$

(e)   Square each score in all conditions and sum these squared scores:

$$\Sigma X_T^2 = 3^2 + 2^2 + 4^2 + 3^2 + 6^2 + 2^2 + 3^2 + 5^2 + 12^2 + 8^2 + 6^2 + 6^2$$

$$= 9 + 4 + 16 + 9 + 36 + 4 + 9 + 25 + 144 + 64 + 36 + 36$$

$$= 392$$

(f)   Count the scores in each condition:

$$N_1 = 4 \qquad N_2 = 4 \qquad N_3 = 4$$

(g)   Count the total number of scores:

$$N_T = 4 + 4 + 4 = 12$$

(h) Plug appropriate values into formulas for SS_T, SS_B, and SS_W and compute:

$$SS_T = \frac{N_T \Sigma X_T^2 - (\Sigma X_T)^2}{N_T} = \frac{12(392) - 3600}{12} = 92$$

Note: SS_B and SS_W are worked using the unequal N formula from Box 7.2. Either the equal N or the unequal N formula is appropriate for equal N.

$$SS_B = \left[\frac{(\Sigma X_1)^2}{N_1} + \frac{(\Sigma X_2)^2}{N_2} + \frac{(\Sigma X_3)^2}{N_3} \right] - \frac{(\Sigma X_T)^2}{N_T}$$

$$= \frac{144}{4} + \frac{256}{4} + \frac{1024}{4} - \frac{3600}{12}$$

$$= 56$$

Note: The value before the minus sign is used in finding SS_W.

$$SS_W = \Sigma X_T^2 - \left[\frac{(\Sigma X_1)^2}{N_1} + \frac{(\Sigma X_2)^2}{N_2} + \frac{(\Sigma X_3)^2}{N_3} \right]$$

$$= 392 - 356$$

$$= 36$$

Note that these values for SS_T, SS_B, and SS_W are the same values as we obtained with the definitional formula.

$$SS_T = SS_B + SS_W$$

$$92 = 56 + 36$$

Phase 2   Find df_T, df_B, and df_W.

$$df_T = N_T - 1 = 11$$

$$df_B = k - 1 = 2$$

$$df_W = k(N_c - 1) = 3(4 - 1) = 9$$

$$df_T = df_B + df_W$$

Check: $11 = 2 + 9$

Phase 3   Compute F.

$$MS_B = \frac{SS_B}{df_B} = \frac{56}{2} = 28$$

$$MS_W = \frac{SS_W}{df_W} = \frac{36}{9} = 4.0$$

$$F = \frac{MS_B}{MS_W} = \frac{28}{4.0} = 7.0$$

Phase 4 Compare computed F with critical F.

From Table D, the critical F for 2,9 df is 4.26. (Be sure to check this yourself.) Since the computed F in our example is larger than our critical F (i.e., 7.0 > 4.26), we reject H_0 and infer from our experimental design that our independent variable was creating differences between the conditions. A typical way of reporting a significant F is the following: ''A one-factor analysis of variance comparing the three caffeine conditions was significant, $F(2,9) = 7.0$, $p < .05$.'' In this statement, *significant* means that H_0 was rejected, the values in parentheses after F refer to the degrees of freedom in the numerator and denominator of the computed F, the value 7.0 is the computed F value, p is the computed probability of the results occurring under conditions where H_0 is true, and the last value (e.g., .05) represents an α level. It is also common for the results of an ANOVA to be presented in summary form as shown in Table 7.2.

TABLE 7.2
Analysis of Variance Summary Table for Caffeine Study

Source of Variation	SS	df	MS	F	p
Total	92	11			
Between conditions	56	2	28.0	7.0	<.05
Within conditions	36	9	4.0		

Notice that when we reject H_0, we reject the H_0 statement, which is $H_0 = \mu_1 = \mu_2 = \cdots = \mu_K$. We can thus be confident that all means are not equal when we reject H_0. But the ANOVA is a general test. It does not allow us to say which specific pairs of means are different from one another. To answer this question (assuming we have more than two conditions in our experiment), we will have to do follow-up tests.

MULTIPLE COMPARISONS: THE PROTECTED t TEST

Except in the simplest situations, the ANOVA can be regarded as a general test to determine whether something happened in an experiment. If the *F* test is nonsignificant, the usual procedure is not to make further comparisons between sets of conditions within the experiment. However, if the *F* test is significant (i.e., H_0 is rejected), then usually the investigator wants to know which means are significantly different from one another. For example, in our caffeine experiment, a significant *F* does not, by itself, tell us specifically what happened. To illustrate, the one-pill condition may have had no effect at all on the task, or it may have had an effect equivalent to the three-pill condition, or perhaps the effect was somewhere in between. It could even have had an effect greater than the three-pill condition or less than the placebo condition, although that type of outcome would not be likely. *To fully interpret the effects of a significant ANOVA (having more than two conditions), the investigator should*

(a) Display the means of the conditions in a table or graph so that the direction and magnitude of differences between the means of the conditions become apparent.
(b) Perform statistical tests on specific comparisons between means of conditions that are of interest.

With respect to the latter, we could conceivably do three paired comparison tests with our caffeine study: placebo versus one pill, placebo versus three pills, one pill versus three pills. If we make these comparisons with three *t* tests (or three two-condition ANOVAS), two problems arise. First, since the same data are being used more than once within the three tests, the tests are not independent. Second, the more tests that we perform, the greater the likelihood of our making a Type I error (falsely rejecting H_0). The probability of making a Type I error with three sets can be as much as three times the probability of making a Type I error for one test. This is true for the same reason that buying three lottery tickets gives one three times the chance of winning as does buying one ticket. Since there are $(k-1) + (k-2) + \cdots + (k-k)$ paired comparisons for *k* conditions, with five conditions there would be 10 (i.e., $4 + 3 + 2 + 1$) possible paired comparisons. For eight conditions, there would be 28 paired comparisons, and so on. The point is that with numerous tests, the probability becomes quite high that the investigator will make one or more Type I errors. That is, the experimenter will claim that something happened because of an independent variable when actually it was a random event. The reader is referred to more advanced statistic tests for a more in-depth discussion of these problems and solutions. References are given at the end of the chapter.

post-hoc tests:

tests done on specific comparisons following a general test such as an analysis of variance

In reading psychology literature, you will come across several types of *post-hoc tests*. Among the common ones are the Tukey, Scheffé, Duncan, Newman-Kuels, and Fisher. These tests vary somewhat in how conservative they are with respect to making α errors.

**protected *t*
test:**

t tests performed following a significant *F*
in an analysis of variance

We will illustrate the Fisher LSD (least significant difference) procedure, also known as ***protected t tests*** for two reasons. First, simulations (termed *Monte Carlo* studies) have found that this type of post-hoc test does very well in terms of making correct decisions concerning rejecting or retaining H_0 (Carmer and Swanson, 1973). Second, they are based on a distribution that you are already familiar with—the *t* distribution. These tests are "protected" (i.e., reducing the probability) from Type I error by using them *only if* the overall ANOVA is significant. These tests also use MS_W in the error term (denominator) of the *t* tests instead of the pooled variance normally used in a *t* test. The tests are easy to compute and apply. They are described next.

WHEN THE *N*'s IN ALL CONDITIONS ARE EQUAL

(a) Find LSD. Any two means from the experiment that have an absolute (i.e., ignore the sign) difference (e.g., $|\bar{X}_1 - \bar{X}_2|$) larger than the LSD value will be significant at the chosen α level:

$$LSD = t\sqrt{MS_W\left[\frac{2}{N_c}\right]}$$

where t = critical t from the t table with df_W as
degrees of freedom for the chosen α
level for a nondirectional test
$MS_W = MS_W$ from ANOVA

N_c = number of scores per condition

(b) Compare the difference between any or all pairs of means with the LSD value. If the absolute difference between any pair of means is greater than the LSD value, it is significant at the chosen α level. Any absolute difference between means less than the LSD value is not significant at the chosen α level.

UNEQUAL *N*'s

For unequal *N*'s per condition, each paired comparison test must be made separately by using the formula

$$t = \frac{\bar{X}_i - \bar{X}_j}{\sqrt{MS_W\left[\frac{1}{N_i} + \frac{1}{N_j}\right]}}$$

where t = computed t value
$\bar{X}_i, \bar{X}_j$ = any two means
$MS_W = MS_W$ from ANOVA
N_i, N_j = number of scores in conditions for which
the comparison is made

This computed t value is then compared with the critical t value from a t table, using df_W as degrees of freedom in conjunction with the chosen α level for a nondirectional test.

LSD FOR CAFFEINE EXPERIMENT

For our caffeine experiment we have four subjects per condition (i.e., equal N), so we can make multiple comparisons using LSD. We need the following values:

(a) α level: we will use the .05 level.

(b) df_W: this was 9 for our experiment.

(c) MS_W: this was 4.0 for our experiment.

(d) N_c: this was 4 for our experiment.

(e)   critical t: from Table C using df_W (i.e., 9) and $\alpha = .05$ for a nondirectional test, the critical t value is 2.26.

Therefore,

$$LSD = 2.26\sqrt{4.0\left[\frac{2}{4}\right]} = 3.20$$

Because any difference between pairs of means greater than 3.20 is significant at $\alpha = .05$, the difference between the means of the placebo and the one-pill conditions (i.e., three versus four) was not significant, whereas the difference between the placebo and the three-pill (i.e., three versus eight) conditions and the difference between the one-pill and three-pill (i.e., four versus eight) conditions were significant. When tests for several comparisons are made, they should be presented in a table for the reader's convenience.

We will illustrate the unequal N procedure by comparing the placebo-versus-one-pill conditions in our study. The unequal N procedure is also appropriate for equal N's. The advantage of the LSD format for equal N's is the ease with which several comparisons can be made.

With a nondirectional test with $\alpha = .05$ and 9 df (the number of df in df_W), the critical t value is 2.26. Using

$$t = \frac{\bar{X}_1 - \bar{X}_2}{\sqrt{MS_W\left[\frac{1}{N_1} + \frac{1}{N_2}\right]}}$$

we have

$$t = \frac{4 - 3}{\sqrt{4.0(1/4 + 1/4)}} = .71$$

Since $.71 < 2.26$, retain H_0. (H_0: $\mu_{placebo} = \mu_{1\,pill}$)

ANOVA AND EXPERIMENTAL DESIGN

between-subject design:

design where different groups of subjects participate in the conditions representing the independent variable or variables

within-subject design:

same as repeated-measures design

Recall that there were two types of *t* tests: one appropriate for independent groups and one appropriate for dependent groups. A similar situation is true for ANOVA designs. With ANOVA, *independent groups (different subjects in each condition) are commonly called **between-subject designs**. Dependent groups are commonly called **within-subject designs**. Within-subject designs are also called **repeated-measures** designs, since matching is typically done by having each subject participate in all conditions.* The type of dependent groups test called matched groups is rarely used with more than two conditions. The same advantages and disadvantages that were discussed for dependent-versus-independent groups *t* tests hold for their ANOVA counterparts. This chapter has dealt only with the between-subject situation. A corresponding within-subject design will be developed in Chapter 9. In Chapter 8 we will show how the principles that we have developed for the one-factor situation can be applied to a more complicated situation involving two factors.

Box 7.3
Computational Formula for Finding *F* Using Calculators That Automatically Compute Standard Deviations

---

Note. With the type of calculator having an S key, you can compute an *F* value in from one third to one tenth of the time it would take you with more typical calculators. The following procedures are appropriate for either equal *N* or unequal *N* situations.

Using the S key

a   Find SS for each of the conditions in the experiment by squaring each S and multiplying by $N_c - 1$.

b   Since $MS_W = (SS_1 + SS_2 + \cdots + SS_k)/df_W$, we can find MS_W by summing SS values (from step 1) across the conditions to obtain SS_W. Then we divide SS_W by df_W.

c   Find SS_T by finding S, using all scores in all conditions as one group of scores.

$$SS_T = (N_T - 1)S^2$$

that is, square the S value and multiply by $N_T - 1$.

d   Subtract SS_W from SS_T to get SS_B.

e   Find MS_B by dividing the value in step 4 (SS_B) by $k - 1$ (the number of conditions minus 1).

f $F = MS_B$ (the value in step 5)/MS_W (the value in step 2)

We will demonstrate these steps, using the data from our caffeine experiment. The data were

Box 7.3 (Continued)

Placebo	One Pill	Three Pills
3	6	12
2	2	8
4	3	6
3	5	6

Steps

a SS (placebo) = 3(.67) = 2.00
   SS (1 pill) = 3(3.33) = 10.00
   SS (3 pills) = 3(8.00) = 24.00

b MS_W = (2.00 + 10.00 + 24.00)/9.00 = 36.00/9.00 = 4.00

c SS_T = 11(8.36) = 92.00

d SS_B = 92.00 − 36.00 = 56.00

e MS_B = 56/2 = 28.00

f F = 28/4 = 7.00

BOX 7.4
Summary of Symbols

Symbol	Meaning
N_T	Total number of scores
N_c	Number of scores in a condition
k	Number of conditions; also the last condition in a series (i.e., X_1, X_2, $\cdots$, X_k)
$\bar{X}_G$	The grand mean—the mean of all scores in all conditions
$\bar{X}_c$	Mean of scores in a condition
X	A score
X with subscript	Score in the condition represented by the subscript
SS	Sum of squares [i.e., $\Sigma (X - \bar{X})^2$]
MS	Mean squares (i.e., SS/df)
MS_B	Mean squares between groups (i.e., SS_B/df_B)
MS_W	Mean squares within groups (i.e., SS_W/df_W)
F	The ratio MS_B/MS_W

SUMMARY

A one-factor analysis of variance (ANOVA) utilizes the F distribution to test for significant differences between two or more means. These means represent data from conditions having one factor (i.e., one independent variable). The F test consists of the following:

$$F = \frac{MS_B}{MS_W}$$

where mean square between (MS_B) represents variability (i.e., differences) between the means of conditions, and MS_W represents variability within the conditions. The same logic for testing hypotheses that was developed using the t test is also used in the ANOVA.

The components of the computed F value in the one-factor ANOVA are

$$F = \frac{MS_B}{MS_W}$$

$$MS_B = \frac{SS_B}{df_B}$$

$$MS_W = \frac{SS_W}{df_W}$$

It can be shown that the SS breakdown in the one-factor ANOVA is

$$SS_T = SS_B + SS_W$$

where SS_T is the variability when all scores are considered as one group of scores, SS_B is the variability between the conditions weighted (i.e., multiplied) by the number of scores per condition, and SS_W is the summed variability within conditions. Similarly,

$$df_T = df_B + df_W$$

where df_T is the total degrees of freedom, df_B is the degrees of freedom between conditions, and df_W is the summed degrees of freedom within conditions.

When the null hypothesis is true, the most likely computed F values are those close to 1. The larger the treatment effects (i.e., differences between the means of the conditions of the independent variable), the larger will be the computed F value. A computed F value must be greater than 1 before statistical significance can be achieved.

The F distributions are a family of mathematical curves having two parameters. One finds a critical F from an F table by using df_B to find the appropriate column and df_W to find the appropriate row. All of the rejection region for H_0 is within the tail of an F distribution. Properties of F distributions are

(a) F values range from zero to positive infinity.
(b) F distributions are *not* symmetrical, but are skewed right.
(c)   The F distributions become more normal in shape as the values of both parameters become larger.
(d)   The F distributions get increasingly closer to the x axis with values greater than 1.

The same four steps that were used to make statistical inferences in the t test are used in the ANOVA. They are

(a)   Determine the α level and the degrees of freedom in MS_B and MS_W.
(b)   From a table containing critical F values, find the appropriate critical F by finding the column represented by df_B and the row represented by df_W. Most tables, including Table D in the back of this book, will give at least two critical F values for each $df_B - df_W$ combination, representing levels of .05, .01, and perhaps others.
(c)   Find the computed F value.
(d)   If the computed F value is larger than the critical F value, reject H_0. If not, retain H_0, where H_0 is defined as ''H_0 represents $\mu_1 = \mu_2 = \cdots = \mu_k$.''

The ANOVA is a general test. When H_0 is rejected, it does not allow us to say which specific pairs of means are different from one another. One solution to this problem is to use *Fisher protected t tests*. These tests are ''protected'' against Type I errors by using them only if the F for the ANOVA is significant (i.e., H_0 is rejected). If the N within each condition of the independent variable is the same, a value called LSD can be used, from which all pairs of means can be tested for statistically significant differences. The LSD value is

$$LSD = t\sqrt{MS_W \left[\frac{2}{N_c} \right]}$$

where t = the critical t from the t table with df_W as
degrees of freedom and using the
chosen α level for a nondirectional test

$$MS_W = MS_W \text{ from ANOVA}$$

$$N_c = \text{number of scores per condition}$$

Absolute differences between any pair of means that exceed the LSD value are significant at the chosen α level. If there are unequal N's per condition, each pair of means must be tested separately by using the formula

$$t = \frac{\bar{X}_1 - \bar{X}_2}{\sqrt{MS_W \left[\dfrac{1}{N_i} + \dfrac{1}{N_j} \right]}}$$

KEY TERMS

t test

F test

F distribution

ANOVA

SS breakdown in ANOVA

"error" term

post-hoc tests

protected t tests

Verbally state how to compute
  the following using the

definitional formula (where
appropriate).

SS_T

SS_B

SS_W

df_T

df_B

df_W

computed F

critical F

PROBLEMS

[Note: Use $\alpha = .05$ to test H_0 in all problems. Where appropriate, follow the four-step outline given in the chapter. Put the results of all ANOVAs in summary tables with the critical F below the summary table.]

1. On the following set of data, do an independent groups t test (from Chapter 6) and an ANOVA. Does $t^2 = F$ for the computed values? Does $t^2 = F$ for the critical values found in the tables?

A	B
14	8
10	5
5	7
8	6
13	4

2. Use the ANOVA definitional formula to see if significant differences exist between the means of the following three conditions:

A	B	C
31	33	29
28	27	35
29	29	35
28	31	33

3. Do problem 2 using the computational formula.

4. Add 9 units to each number in the C column in problem 2 and replace the old numbers with these new numbers. Then redo the ANOVA and put the results in a summary table. Why did SS_B change values (compared with problem 2), but SS_W did not?

5. Suppose we add 12 units to the first number in each of the three conditions (A, B, C) in problem 2, replace the old numbers with the new, and redo the ANOVA. How would the new SS_B compare in size with the old SS_B? Why? How would the new SS_W compare in size to the old SS_W? Why? If you are sure of your answers, you need not work out the new ANOVA. If you are not sure of your answers, work out the new ANOVA and answer the questions.

6. A researcher was interested in the running speed of rats as a function of schedule of reinforcement, following 10 continuously reinforced trials. Four schedules were used: 1:1, 2:1, 4:1, 8:1. Running speed was measured in terms of average time in seconds from the start box to the goal box for 16 trials. Did significant differences occur as a function of schedules of reinforcement? If so, where? (Do an ANOVA and protected t tests if necessary.)

1:1	2:1	4:1	8:1
2.1	1.6	2.2	2.8
2.4	1.9	2.5	2.9
1.8	1.8	2.0	2.9
2.0	1.7	2.1	2.8
2.2	1.5	2.2	2.6

7. An investigator wishes to know if the short-term memory span is different for random strings of letters, numbers, or one-syllable

words. The dependent variable is the number of correctly sequenced items before a mistake is made. Did significant differences exist among the conditions? If so, where? (Note the N's in the conditions.) State the results of the ANOVA as you might find it written in a journal article.

Letters	Numbers	Words
8	9	8
7	7	7
7	8	7
6	8	8
9	9	7
8	7	6
7	7	7
8	8	7
8		8

8. In a brainstorming experiment, an investigator wants to know if the size of a group will affect the number of creative solutions given to a problem. Three group sizes were used: 1, 3, 6. Did the groups significantly differ? If so, which groups were better? Write the results of the ANOVA as you might find them written in a journal article.

Group Size

1	3	6
8	9	21
10	12	19
13	18	20
4	15	23
15	20	17
9	19	18
	16	
	14	

9. An investigator wants to test the effects of position within a sentence (first, middle, last) on the detection of spelling errors. Three groups of subjects each read a passage where 10 spelling errors were in either the first, middle, or last position of sentences. Were there significant differences between positions? If so, which conditions were significantly different?

First	Middle	Last
8	4	9
6	3	6
9	5	5
10	4	7
7	6	8
5	3	8
8	1	6
7	4	8

10. An investigator did a one-factor ANOVA on five conditions (A, B, C, D, E); 10 subjects were in each condition. The computed F was statistically significant, with $MS_W = 5.41$. The cell means for the five conditions were $A = 54$, $B = 45.6$, $C = 51.7$, $D = 49.6$, and $E = 47.8$. Find the appropriate LSD value, test all pairs of means, and list the significant comparisons by using the following format:

$$A > B, \quad p < .05$$
$$A > E, \quad p < .05$$
$$\text{etc.}$$

11. *a* Do a protected t test on the data in problem 1.

   b How does the size of the computed t value in the protected t test compare with the computed t value in the regular t test that you did in problem 1?

   c How does MS_W from an ANOVA having two conditions compare to the pooled variance (S^2_{pooled}) in a t test?

12. Using three conditions with four scores per condition, make up one set of data such that the following constraints are met:

a An ANOVA on these data would be significant.

b The grand mean is 10.

c The condition means are whole numbers.

d The numbers within the conditions are whole numbers and are not all the same.

   I. Do the ANOVA and put the results in a summary table.

   II. Given that $\bar{X}_G = 10$, how many condition means were free to vary (i.e., they could be any number)? How does this compare with df_B? Given the condition means, how many scores within each condition were free to vary? Add the number of scores that are free to vary within each condition. How does this number compare with df_W?

REFERENCES

Comparisons of selected post-hoc tests can be found in the following books.

Edwards, A. L. (1972). *Experimental design in psychological research* (4th ed.). Fort Worth, Texas: Ch. 8.

Kirk, R. E. (1968). *Experimental design: Procedures for the behavioral sciences.* Belmont, California: Brooks Cole, Ch. 3.

The following article shows the protected *t* test to perform well as a post-hoc test.

Carmer, S. G., and Swanson, M. R. (1973). An evaluation of ten multiple comparison procedures by Monte Carlo methods. *Journal of the American Statistical Association, 68,* 66–74.

CHAPTER 8

Two-Factor Analysis of Variance

INTRODUCTION

interaction:

the differential effect on a dependent variable coming from different combinations of conditions of two or more independent variables

Thus far, we have considered designs having only one independent variable. These designs are quite appropriate for situations where the interest in a topic does not go beyond how the specific conditions within an independent variable affect the dependent variable. For example, one might be interested in which of three textbooks is best for teaching spelling to the second graders within a particular school district. However, in many situations, an independent variable may operate in conjunction with another independent variable to produce effects on the dependent variable that would not occur with either independent variable alone. This occurrence is called an *interaction*. An *interaction is thus defined as the differential effect on a dependent variable coming from different combinations of conditions of two or more independent variables.* Interactions may occur with any number of independent variables (i.e., factors), but the focus of this chapter is on two independent variables. The way in which one determines whether an interaction exists in a two-factor experiment is to do an experiment where conditions of one independent variable are combined (i.e., crossed) with conditions of a second independent variable. Their combined effects on a dependent variable can then be measured. *Experimental designs joining all levels of two or more independent variables (i.e., factors) to test their effect on a dependent variable are called **factorial designs.** A by-product of a factorial design is that when no significant interaction exists, each factor may be interpreted as having independently affected the dependent variable.* In this situation, the two-factor experiment becomes essentially two one-factor experiments.

RELATIONSHIP BETWEEN ONE- AND TWO-FACTOR DESIGNS

The simplest way to understand a two-factor design is to think of it as providing three separate tests on a set of data. First, independent variable *A* is tested, assuming independent variable *B* did not exist. Then, independent variable *B* is tested, assuming independent variable *A* did not exist. The last test is for joint effects of *A* and *B* (i.e., the interaction). Let us show how all of this works by presenting an example with the appropriate analysis.

Mr. Smith is a researcher who is interested in the amount of talking done by clients during therapeutic sessions. He hypothesizes that the gender of the therapist will make a difference on the amount of talking done by clients. He, therefore, designs an experiment to test his hypothesis. He first finds a male therapist and a female therapist who are matched on all variables that might affect talking by clients, except, of course, gender. (This, in practice, would be very difficult to do, but for simplicity we will make this assumption.) He then randomly selects 12 clients and randomly assigns them to the two therapists. He then records the number of minutes that each client spent talking during a 30-minute session. Although he noted the gender of the client on his data sheet, he ignored this in his analysis. Since there is one independent variable (i.e., factor) having two conditions (i.e., male versus female therapist), and since there are different subjects in each condition, we have a one-factor, between-subject design containing two conditions. In this situation, either a *t* test for independent groups or a between-subject one-factor ANOVA is appropriate. Since our current topic is ANOVA, let us analyze the data using a one-factor ANOVA. The data with gender of the client noted beside them is shown in Table 8.1.

TABLE 8.1
Minutes Spend Talking by Clients as a Function of the Gender of the Therapist
(hypothetical data)

Gender of Therapist	
Male	Female
11 F	16 M
21 M	23 F
12 F	12 M
20 M	20 F
19 M	14 M
13 F	23 F
($\bar{X} = 16$)	($\bar{X} = 18$) ($\bar{X}_G = 17$)

A one-factor ANOVA on this data showed the following (using the definitional formula):

$$SS_T = \Sigma(X - \bar{X}_G)^2 = (11-17)^2 + (21-17)^2 + (12-17)^2 + (20-17)^2$$
$$+ (19-17)^2 + (13-17)^2 + (16-17)^2 + (23-17)^2 + (12-17)^2$$
$$+ (20-17)^2 + (14-17)^2 + (23-17)^2 = 222$$

$$SS_B = N_c[(\Sigma\bar{X}_c - \bar{X}_G)^2] = 6[(16-17)^2 + (18-17)^2] = 6(2) = 12$$

$$SS_W = \Sigma(X - \bar{X}_c)^2 = (11-16)^2 + (21-16)^2 + (12-16)^2 + (20-16)^2$$
$$+ (19-16)^2 + (13-16)^2 + (16-18)^2 + (23-18)^2$$
$$+ (12-18)^2 + (20-18)^2 + (14-18)^2 + (23-18)^2 = 210$$

$$df_B = k - 1 = 2 - 1 = 1$$

$$df_W = k(N_c - 1) = 2(5) = 10$$

$$df_T = N_T - 1 = 11$$

$$MS_B = \frac{12}{1} = 12$$

$$MS_W = \frac{210}{10} = 21$$

$$F = \frac{12}{21} = .57$$

The critical $F_{1,10}$ ($\alpha = .05$) = 4.96 (from Table D). Since .57 < 4.96, we retain H_0. Mr. Smith therefore concludes in his study that the gender of the therapist did not make a difference in amount of talking by clients.

Ms. Jones is interested in gender differences in quantity of talking by clients in therapeutic situations. Upon looking at Mr. Smith's data, she realizes that if the data were grouped in a different way, a test could be done to see if there are gender differences in clients. She therefore regroups the data by putting all female clients in one group and all male clients in a second group. Again, we have a between-subject one-factor design. When regrouped, the data are as in Table 8.2.

TABLE 8.2

Gender of Client	
Male	Female
21	11
20	12
19	13
16	23
12	20
14	23
$\bar{X} = 17$	$\bar{X} = 17$ $\bar{X}_G = 17$

Ms. Jones then performed a one-factor ANOVA on the data as follows:

$$SS_T = (21-17)^2 + (20-17)^2 + (19-17)^2 + (16-17)^2 + (12-17)^2 + (14-17)^2$$
$$+ (11-17)^2 + (12-17)^2 + (13-17)^2 + (23-17)^2 + (20-17)^2 + (23-17)^2$$
$$= 222$$

$$SS_B = 6[(17-17)^2 + (17-17)^2)] = 0$$

$$SS_W = (21-17)^2 + (20-17)^2 + (19-17)^2 + (16-17)^2 + (12-17)^2 + (14-17)^2$$
$$+ (11-17)^2 + (12-17)^2 + (13-17)^2 + (23-17)^2 + (20-17)^2 + (23-17$$
$$= 222$$

Note that since each condition mean is the same as the grand mean (i.e., 17), $SS_B = 0$ and $SS_W = SS_T$.

$$df_B = 2 - 1 = 1$$

$$df_W = 5(2) = 10$$

$$df_T = 12 - 1 = 11$$

$$MS_B = \frac{0}{1} = 0$$

$$MS_W = \frac{222}{11} = 20.18$$

$$F = \frac{0}{20.18} = 0$$

Critical $F_{1,10} = 4.96$. Since $0 < 4.96$, retain H_0.

Because of the results of this test, Ms. Jones concludes that the amount of talking by clients is not affected by the gender of the client. Based on the results of the test on therapist gender and the test on client gender, it appears as though gender is not an important factor in the speech productivity of clients. Enter now Statman, who has just completed a course on two-factor ANOVA. He speculates that perhaps there is a joint effect of gender of therapist and gender of client on a client's speech productivity. In other words, specific combinations of gender of therapist and gender of client may occur that do make a difference in the speech productivity of a client. Therefore, Statman rearranges the data a third time. This time, two factors are created (i.e., gender of therapist and gender of client) with two conditions within each factor (i.e., male and female). The data would be arranged as in Table 8.3.

This arrangement contains four specific combinations of the two independent variable conditions: male client, male therapist; male

TABLE 8.3
Two-Factor Arrangement of Data in Example

		Gender of Client		
		Male	Female	
Gender of Therapist	Male	21 20 $\bar{X} = 20$ 19	11 12 $\bar{X} = 12$ 13	$\bar{X}_M = 16$
	Female	**16** 12 $\bar{X} = 14$ 14	**23** 20 $\bar{X} = 22$ 23	$\bar{X}_F = 18$
		$\bar{X}_M = 17$	$\bar{X}_F = 17$	$\bar{X}_G = 17$

client, female therapist; female client, male therapist; female client, female therapist. *These combinations of independent variable conditions are called **cells.*** If one looks at the four cell means from our example, a different picture emerges than from that obtained by looking at each independent variable separately. It now appears that when the genders of the client and therapist are the same, much more talking is elicited than when the genders are different. Let us now outline a procedure for doing a two-factor ANOVA, and apply this procedure to our example.

OUTLINE FOR TWO-FACTOR ANOVA

When the conditions or levels of two factors are crossed to form a matrix, such as the one in Table 8.3, a series of rows and columns will result. In our example, there are two rows and two columns. Designs are often identified by listing the number of rows by the number of columns. Thus, our design is a 2×2 design (read as 2 by 2). If we had two rows and three columns, it would be called a 2×3 design, etc.

A two-factor ANOVA will arrange the data in three separate, independent ways to give three separate statistical tests using the F distribution as a model. One value, F_r, is used in testing for differences between row means. A second value, F_{col}, is used in testing for differences in column means. *Differences between row means or between column means are often termed **main effects.*** The term *main* does not mean and should not be confused with *major*. For example, one might read an article wherein the main effect for rows was reported as nonsignificant, whereas the main effect for columns was reported as being significant. The third value, F_I, is used in testing for interaction (i.e., differential effect on the dependent variable coming from different combinations of independent variable conditions) between the two independent variables. When row means are tested, the column distinctions are ignored. In similar fashion, when column means are tested, row distinctions are ignored. However, data from cells representing combinations of specific rows and columns are evaluated in the test for the interaction.

main effects:

differences between condition means within an independent variable (e.g., between row means or between column means) in an analysis of variance

It can be shown that for a two-factor ANOVA

$$SS_T = SS_r + SS_{col} + SS_I + SS_W$$

(The total SS can be broken down into SS for rows, SS for columns, SS for interaction, and SS within cells.) Also

$$SS_T = \Sigma(X - \bar{X}_G)^2$$

Each score minus the grand mean squared and summed.

$$SS_r = N_r\Sigma(\bar{X}_r - \bar{X}_G)^2$$

Each row mean minus the grand mean squared and summed. Then this sum is multipled by the number of scores in a row.

$$SS_{col} = N_{col}\Sigma(\bar{X}_{col} - \bar{X}_G)^2$$

Each column mean minus the grand mean squared and summed. Then this sum is multiplied by the number of scores in a column.

$$SS_W = \Sigma(X - \bar{X}_c)^2$$

Each score minus the mean of the cell that contains the score is squared and summed.

$$SS_i = N_c\Sigma(\bar{X}_c - \hat{X}_c)^2$$

Each cell mean minus the predicted cell mean if there were no interaction, squared, summed, then multiplied by the number of scores in a cell.

Since finding predicted cell means (i.e., $\hat{X}$) under conditions of no interaction involves some new concepts, we will delay the presentation of finding SS_I through the definitional formula given. Instead, we will find SS_I by solving the equation $SS_T = SS_r + SS_{col} + SS_I + SS_W$ for SS_I. We therefore have $SS_I = SS_T - SS_r - SS_{col} - SS_W$. Repeating the data from our example, we have

Gender of Client

		Male	Female	
Gender of Therapist	Male	21 20 $\bar{X}_{cell}$ = 20 19	11 12 $\bar{X}_{cell}$ = 12 13	$\bar{X}_M$ = 16
	Female	16 12 $\bar{X}_{cell}$ = 14 14	23 20 $\bar{X}_{cell}$ = 22 23	$\bar{X}_F$ = 18
		$\bar{X}_M$ = 17	$\bar{X}_F$ = 17	$\bar{X}_G$ = 17

$$SS_T = (21-17)^2 + (20-17)^2 + (19-17)^2 + (16-17)^2 + (12-17)^2 + (14-17)^2$$
$$+ (11-17)^2 + (12-17)^2 + (13-17)^2 + (23-17)^2 + (20-17)^2 + (23-17)^2$$
$$= 222$$
$$SS_r = 6[(16-17)^2 + (18-17)^2] = 12$$

$$SS_{col} = 6[(17-17)^2 + (17-17)^2] = 0$$

$$SS_W = (21-20)^2 + (20-20)^2 + (19-20)^2 + (16-14)^2 + (12-14)^2 + (14-14)^2$$
$$+ (11-12)^2 + (12-12)^2 + (13-12)^2 + (23-22)^2 + (20-22)^2 + (23-22)^2$$
$$= 18$$

$$SS_I = 222 - 12 - 0 - 18 = 192$$

These are the same SS values that we obtained with the computational formulas given in Box 8.1.

df BREAKDOWN FOR TWO-FACTOR ANOVA

$$df_T = df_r + df_{col} + df_I + df_W$$

Degrees of freedom can be broken down into components in the form of a "degrees of freedom tree" shown in Illustration 8.1. The exact same tree can be made for SS by substituting SS for df.

Illustration 8.1

Partitioning Degrees of Freedom with Number for Our Example in Parentheses

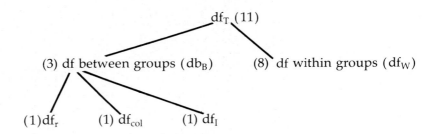

The formulas for finding these dfs are as follows:

$df_T = N_T - 1$	Total number of scores minus 1.
(df between groups = number of cells − 1)	
$df_r = r - 1$	Number of rows minus 1.
$df_{col} = col - 1$	Number of columns minus 1.
$df_W = (r \times col)(N_c - 1)$	Number of scores in a cell minus 1 times the number of cells. The number of cells is $r \times col$.
$df_I = (r - 1)(col - 1)$	Number of rows minus 1 times the number of columns minus 1.

The dfs for our example would be

$df_T = 12 - 1 = 11$

df between $= 4 - 1 = 3$

$df_r = 2 - 1 = 1$

$df_{col} = 2 - 1 = 1$

$df_I = (2 - 1)(2 - 1) = 1$

$df_W = (2 \times 2)(3 - 1) = 8$

Check: $11 = 3 + 8; \quad 3 = 1 + 1 + 1$

MEAN SQUARES (MS) AND F's FOR TWO-FACTOR ANOVA

$$MS_r = \frac{SS_r}{df_r}$$

$$MS_{col} = \frac{SS_{col}}{df_{col}}$$

$$MS_W = \frac{SS_W}{df_W}$$

$$MS_I = \frac{SS_I}{df_I}$$

$$F_r = \frac{MS_r}{MS_W}$$

$$F_{col} = \frac{MS_{col}}{MS_W}$$

$$F_I = \frac{MS_I}{MS_W}$$

Note. **MS$_W$ is called the "error term" and will be the denominator of the F tests.**

Using our example, the corresponding values are

$$MS_r = \frac{12}{1} = 12$$

$$MS_{col} = \frac{0}{1} = 0$$

$$MS_W = \frac{18}{8} = 2.25$$

$$MS_I = \frac{192}{1} = 192$$

$$F_r = \frac{12}{2.25} = 5.33 \quad \text{critical } F_{1,8} \, (\alpha = .05) = 5.32$$

$$F_{col} = \frac{0}{2.25} = 0 \qquad \text{critical } F_{1,8} \ (\alpha = .05) = 5.32$$

$$F_{I} = \frac{192}{2.25} = 85.2 \qquad \text{critical } F_{1,8} \ (\alpha = .05) = 5.32$$

To find the critical F from which to compare the computed F's in order to determine whether to reject H_0, we use Table D in the appendix. The proper value in Table D is found by reading the df in the numerator of the desired F test across the top of the table and by reading the df in the denominator (i.e., error term) down the side of the table. These two degrees of freedom are often listed as subscripts for the critical F. Thus, the critical F for rows in our example using $\alpha = .05$ would be

$$\text{critical } F_{1,8} (\alpha = .05) = 5.32$$

Since the critical F's for columns and interactions in our example also involve 1 and 8 df, the same critical F value is used to test all of the computed F's in our example. This will not usually be the case. Our decision concerning retaining or rejecting H_0 for our three tests would be

For rows (which represent gender of the therapist), $5.33 > 5.32$; therefore reject H_0.	H_0: The population means for rows are the same.
For columns (which represent gender of client), $0 < 5.32$; therefore retain H_0.	H_0: The population means for columns are the same.
For interaction, $85.2 > 5.32$; therefore reject H_0.	H_0: The actual population cell means are equal to the predicted cell means.

INTERPRETING A TWO-FACTOR ANOVA

The key to interpreting the two-factor ANOVA involves the interaction as follows: (a) If the F test for interaction is *not* significant, then the tests for rows and columns can be interpreted with no qualifiers. (b) If the interaction is significant, row and column tests can be misleading if not interpreted in relation to the interaction. In fact, when the F test for interaction is significant, often the F tests for rows and columns are not interpreted because the interpretation is meaningless. (c) The results of the statistical tests should be interpreted in reference to a table or graph of cell means so that the direction of comparisons (i.e., which is largest) and general patterns become apparent.

In our example, the interaction was significant. Different combinations of gender make a big difference in amount of client speech production when the client's speech productivity is being measured. Because of this extreme interaction, the row and column tests are not meaningful by themselves. For example, the F test for columns, representing clients, was nonsignificant. However, the gender of the client was very important when combined with gender of the therapist. Table 8.3, showing cell means, clarifies the nature of the interaction that we obtained. When the therapist-client gender was the same, speech productivity was much greater than when it was different.

ONE-FACTOR ANOVA VERSUS TWO-FACTOR ANOVA

The results for our example from both of the one-factor ANOVAs and the two-factor ANOVA are shown in Tables 8.4–8.6.

TABLE 8.4
Summary Table for One-Factor ANOVA Involving Gender of Therapist

Source of Variation	SS	df	MS	F	p
Between groups	12	1	12	.57	> .05
Within groups	210	10	21		
Total	222	11			

TABLE 8.5
Summary Table for One-Factor ANOVA Involving Gender of Client

Source of Variation	SS	df	MS	F	p
Between groups	0	1	0	0	> .05
Within groups	222	10	22		
Total	222	11			

TABLE 8.6
Summary Table for Two-Factor ANOVA

Source of Variation	SS	df	MS	F	p
Rows (therapists)	12	1	12	5.33	< .05
Columns (clients)	0	1	0	0	> .05
Interaction	192	1	192	85.2	< .05
Error	18	8	2.25		
Total	222	11			

In comparing these three summary tables, notice that the SS, df, and MS for the one-factor ANOVA comparing gender of therapist are identical to the SS, df, and MS for rows (which reflect gender of therapist) in the two-factor ANOVA. In similar fashion, SS, df, and MS in the one-factor ANOVA for gender of client are the same as SS, df, and MS for columns (reflecting gender of client) in the two-factor ANOVA. However, note that the F value is not the same for the one-factor test on gender of therapist compared with the F value for rows in the two-factor ANOVA. (Some difference would also ordinarily be true for the column F and the corresponding one-factor F. They are the same in our example only because of the zero in the numerator.) The difference in F values is due to the different denominator (i.e., error term) between the one-factor test and the two-factor test. The one-factor F had scores for males and females combined in the error term. In our situation, this increased the variability within conditions, which, in turn, made the error term large. Since the error term is the denominator of the F test, the larger the error term, the smaller the computed F value. When large interactions, such as that which occurred in our data, or large row or column effects, do not exist, then the error term for a one-factor ANOVA and a two-factor ANOVA may be similar in value.

Because the use of a factorial design can result in a smaller error term than might occur if either of the independent variables were tested using a one-factor ANOVA, experimenters sometimes design experiments to take advantage of this. For example, an experimenter interested in comparing two problem-solving conditions might include year in school as a second independent variable in the design if subjects in his or her experiment varied on the year-in-school dimension. If year in school relates to problem solving, the experimenter may end up with a more powerful (i.e., better chance of correctly rejecting H_0) test for comparing the problem-solving conditions.

To summarize, when one does a two-factor ANOVA, one is essentially doing a one-factor test using the data in the rows and a second one-factor test using the data in the columns. The unique component of the two-factor ANOVA is the test for an interaction.

INTERACTION

SS Development

Let us now develop SS for interaction and in the process learn how to interpret interactions when data are presented in graphic form. We start with a definition. An **effect** *is the difference between a row mean or a column mean and the grand mean.* We will illustrate by again presenting the data from our example.

Gender of Client

		Male	Female	Row $\bar{X}$	Row effects
Gender of Therapist	Male	21 20 $\hat{X}=16$ 19 $\bar{X}=20$	11 12 $\hat{X}=16$ 13 $\bar{X}=12$	16	−1
	Female	16 12 $\hat{X}=18$ 14 $\bar{X}=14$	23 20 $\hat{X}=18$ 23 $\bar{X}=22$	18	+1
Column $\bar{X}$		17	17	$\bar{X}_G = 17$	
Column effects		0	0		

In our example, the effect for

row 1 is $16-17 = -1$ column 1 is $17-17 = 0$
row 2 is $18-17 = +1$ column 2 is $17-17 = 0$

A predicted cell mean (i.e., $\hat{X}$) under conditions of no interaction equals the grand mean plus the row effect plus the column effect for the particular row and column that contain that cell. Thus, the male client, male therapist cell has a predicted cell mean of $17 - 1 + 0 = 16$; the female client, male therapist cell has $\hat{X} = 17 + 1 + 0 = 18$, and so on. The four $\hat{X}$'s are listed in their proper cells in the matrix. SS_I is defined as

$$SS_I = N_c \Sigma (\bar{X}_c - \hat{X})^2$$

Thus, in our example

$$SS_I = 3[(20-16)^2 + (12-16)^2 + (14-18)^2 + (22-18)^2] = 192$$

Observe that this value is the same as that we obtained from the formula $SS_I = SS_T - SS_r - SS_{col} - SS_W$. Notice also that within each column the difference between the $\hat{X}$'s is the same (i.e., $18 - 16 = 2$ and $18 - 16 = 2$). This will always be the case and makes for an easy determination of interaction effects when the cell means are plotted in a graph as shown in Figure 8.1. Notice that the lines are parallel. Now look at a graph (Figure 8.2) of the actual cell means from our example. The graph showing the actual cell means is distinctly nonparallel. The general rule relating interactions to graphs is the following: **The magnitude of interaction increases as a direct function of the degree to which the lines within a graph are nonparallel.**

Let us repeat our interaction story by working another fictitious example. We have a 2[male (M) vs. female (F)] × 3[degree of vio-

Figure 8.1
Predicted cell means if no interaction for our example.

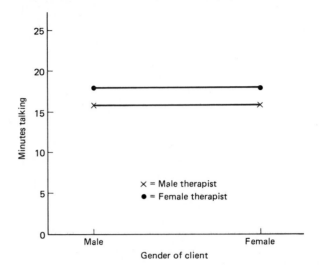

Figure 8.2
Cell means for our example.

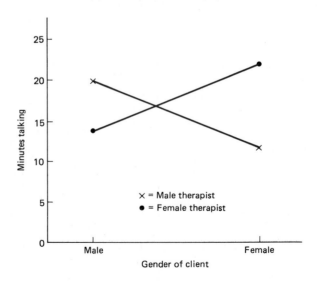

lence: neutral (N), moderate (M), and high (H) displayed in photo-graphs] design. There are 10 people in each cell. The dependent variable is magnified pupil size in millimeters (mm) when looking at the pictures. The cell means and $\hat{X}$'s for the experiment are given in the following matrix:

Type of Picture

		N	M	H	Row $\bar{X}$	Row effects
	M	$\hat{X} = 10$ $(15-1-4)$ $\bar{X} = 11$	$\hat{X} = 15$ $(15+1-1)$ $\bar{X} = 14$	$\hat{X} = 17$ $(15-1+3)$ $\bar{X} = 17$	14	-1
Gender						
	F	$\hat{X} = 12$ $(15+1-4)$ $\bar{X} = 11$	$\hat{X} = 17$ $(15+1+1)$ $\bar{X} = 18$	$\hat{X} = 19$ $(15+1+3)$ $\bar{X} = 19$	16	$+1$
Column $\bar{X}$'s		11	16	18	$\bar{X}_G = 15$	
Column effects		-4	$+1$	$+3$		

You should verify that each $\hat{X}$ is correct. Now,

$$SS_I = 10[(11-10)^2 + (14-15)^2 + (17-17)^2 + (11-12)^2 + (18-17)^2$$
$$+ (19-19)^2] = 40$$

We cannot do an F test since we do not have the within-cell data from which we can compute an error term.

Figure 8.3 shows graphs of the predicted cell means under conditions of no interaction and the actual cell means from this example.

Figure 8.3
Graphs of predicted and actual cell means.

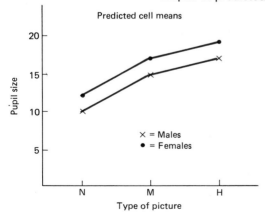

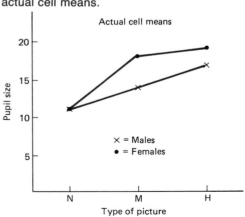

Observe that in the graph of $\hat{X}$'s, males and females react identically to the levels of the other independent variable (i.e., degree of violence in pictures). If the actual data looked like this, we could talk about the effects of the picture without any qualifying remarks concerning gender. Now, notice that in the graph of the actual cell means, a somewhat different interpretation is called for, provided that the F test for interaction is significant. Although each sex has an increased pupil dilation as a function of degree of violence in pictures, the pattern of the increase is different. Females show a larger increase in pupil size going from neutral to moderately violent pictures than do males, whereas males show a larger increase than do females in going from a moderate to a high degree of violence.

The magnitude of an interaction depends on the degree to which the conditions of one independent variable give rise to different patterns across the levels or conditions of a second independent variable. Graphically, these different patterns show up as nonparallel lines. Since different patterns can also occur by chance, the F test for interaction must be significant before different patterns can be interpreted as meaningful. Figure 8.4 shows the shapes of some of the many possible types of interactions.

PROTECTED *t* TESTS
WHEN INTERACTION IS NONSIGNIFICANT

As with a one-factor ANOVA, it is often important and necessary to do follow-up tests between pairs of means to determine where, within a set of data, significant differences lie. Within the constraints of the following rules, all combinations of pairs of means can be tested, and in problem exercises students should do this. However, in published research, usually only the pairs of means that are of specific interest to the researcher are tested. The protected t test that we used following the one-factor ANOVA can also be used following the two-factor ANOVA, but in a somewhat more complicated fashion. The key is the test for an interaction. If this test is nonsignificant, the data are interpreted as though a one-factor ANOVA was performed on each of the two factors. Therefore, when the F test for an interaction is nonsignificant, post-hoc paired comparison tests can be done on row and column means, as described in the chapter on one-factor ANOVA. For example, if in our pupil size study the interaction was nonsignificant, then protected t tests for rows would not be necessary since there are only two rows. However, one might be interested in which of the three column means were significantly different from one another. As with the one-factor situation, we should *only do these follow-up tests if the F test for that factor was significant*. Recall that the formula for finding the LSD (least significant difference) value was

$$LSD = t\sqrt{MS_W\left[\frac{2}{N}\right]}$$

In computing this value to compare row means or column means, MS_W would come from the two-factor ANOVA while N would be the number of scores in a *row* if row means are being compared or the number of scores in a *column* if column means are being compared. The t is the critical t at a chosen α level for the degrees of freedom in MS_W. Therefore, except when there are equal N's in rows and columns, different LSD values will be obtained for rows and columns.

Figure 8.4
Types of interactions.

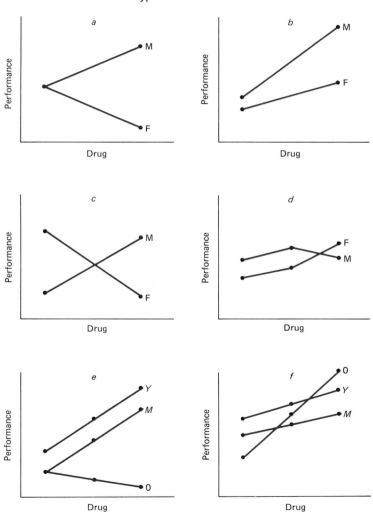

POST-HOC TESTS WHEN INTERACTION IS SIGNIFICANT

If the test for interaction is significant, this indicates that unique patterns exist within the cell means of the design as one looks across rows or columns. Therefore, when the F test for interaction is significant, protected t tests should not be done on differences between row or column means; rather, they should be done to test for significant differences between pairs of cell means. Box 8.1 summarizes the decision-making process for protected t tests. Let us illustrate the procedure for finding LSD by using the data and ANOVA from our example study on gender effects on speech productivity. The cell means were as follows:

Gender of Client

		Male	Female
Gender of Therapist	Male	$\bar{X} = 20$	$\bar{X} = 12$
	Female	$\bar{X} = 14$	$\bar{X} = 22$

There were three scores per cell, so $N = 3$. The MS_W from the ANOVA was 2.25, based on 8 df. Because there were equal N in all cells, we can use LSD to make comparisons between cell means. Using $\alpha = .05$ and 8 df, we find the critical t value from Table C in the appendix to be 2.31. Plugging the appropriate values into the formula for LSD, we have

$$LSD = 2.31 \sqrt{\frac{2.25\,(2)}{3}} = 2.83$$

Thus, the difference between any two cell means that are larger than 2.83 will be significant at $\alpha = .05$. Note, however, that the direction of the differences must be determined from a table or some other listing of cell means. Thus, the male client–female client difference is significant for both male and female therapists, but in opposite directions.

BOX 8.1
Computational Formula for Finding Various SS within a Two-Factor
ANOVA Using the Speech Productivity Example

Step 1   Find SS_T

$$SS_T = \frac{N_T \Sigma X^2 - (\Sigma X)^2}{N_T}$$

where X = each score in the design
N_T = total number of scores in the design

Box 8.1 (Continued)

From Table 8.1, as reproduced here,

Gender of Therapist

		Male	Female
Gender of Client	Male	21 20 19	11 12 13
	Female	16 12 14	23 20 23

$$\Sigma X^2 = 21^2 + 20^2 + 19^2 + 16^2 + 12^2 + 14^2 + 11^2 + 12^2$$
$$+ 13^2 + 23^2 + 20^2 + 23^2$$
$$= 3690$$

$$\Sigma X = 21 + 20 + 19 + 16 + 12 + 14 + 11 + 12 + 13$$
$$+ 23 + 20 + 20 + 23$$
$$= 204$$

$$N_T = 12$$

$$SS_T = \frac{12(3690) - (204)^2}{12} = 222$$

Step 2 Find SS_B by ignoring the factorial design and treating all cells as though they were conditions in a one-factor design.

$$SS_B = \frac{(\Sigma X_{cell\,1})^2}{N_{cell\,1}} + \frac{(\Sigma X_{cell\,2})^2}{N_{cell\,2})} + \cdots + \frac{(\Sigma X_{cell\,k})^2}{N_{cell\,k}} - \frac{(\Sigma X)^2}{N_T}$$

where k is the last cell. Thus, for our example,

$$SS_B = \frac{(21 + 20 + 19)^2}{3} + \frac{(16 + 12 + 14)^2}{3} + \frac{(11 + 12 + 13)^2}{3}$$
$$+ \frac{(23 + 20 + 23)^2}{3} - \frac{(204)^2}{12}$$

$$= 3672 - \frac{(204)^2}{12} = 3672 - 3468$$

$$= 204$$

Step 3 Find SS_W by using the formula $SS_W = SS_T - SS_B$. For our example, $SS_W = 222 - 204 = 18$.

Step 4 Find SS_r by using

$$SS_r = \frac{(\text{Sum of each row})^2}{N_r} - \frac{(\Sigma X)^2}{N_T}$$

Box 8.1 (Continued)

For our example,

$$SS_r = \frac{(21+20+19+11+12+13)^2}{6} + \frac{(16+12+14+23+20+23)^2}{6} - \frac{(\Sigma X)^2}{N_T}$$

$$= 3480 - 3468 \,(\text{as previously computed in step 2})$$

$$= 12$$

Step 5　Find SS_{col} by using

$$SS_{col} = \frac{(\text{sum of each column})^2}{N_{col}} - \frac{(\Sigma X)^2}{N_T}$$

For our example,

$$SS_{col} = \frac{(21+20+19+16+12+14)^2}{6} + \frac{(11+12+13+23+20+23)^2}{6} - \frac{(\Sigma X)^2}{N_T}$$

$$= 3468 - 3468 \,(\text{as previously computed})$$

$$= 0$$

Step 6　Find SS_I by using the formula

$$SS_I = SS_B - SS_r - SS_{col}$$

For our example,

$$SS_I = 204 - 12 - 0 = 192$$

Computational Formula for Calculators That Will Automatically Compute a Standard Deviation

Step 1　Find S by using all scores in all cells as one group.

$$SS_T = (N_T - 1)S^2$$

For our example,

$$SS_T = (11)(20.18) = 222$$

Step 2　First, find S by using the cell means as one group. Then

$$SS_B = N_c[(r \times col) - 1](S^2)$$

(Note. N_c is the number of scores in a cell.) For our example,

$$SS_B = (3)(3)(22.67) = 204$$

Step 3

$$SS_W = SS_T - SS_B$$

For our example,

$$SS_W = 222 - 204 = 18$$

Box 8.1 (Continued)

Step 4   First, find S by using the row means as one group. Then

$$SS_r = N_{row}(r - 1)(S^2)$$

(Note. N_{row} is the number of scores in a row.) For our example,

$$SS_r = 6(1)(2) = 12$$

Step 5   First, find S using the column means as one group, then

$$SS_{col} = N_{col}(c - 1)(S^2)$$

(Note. N_{col} is the number of scores in a column.) For our example,

$$SS_{col} = 6(1)(0) = 0$$

Step 6

$$SS_I = SS_B - SS_r - SS_{col}$$

For our example,

$$SS_I = 204 - 12 - 0 = 192$$

BOX 8.2
Decision-Making Flowchart for Protected *t* Tests for Two-Factor ANOVA

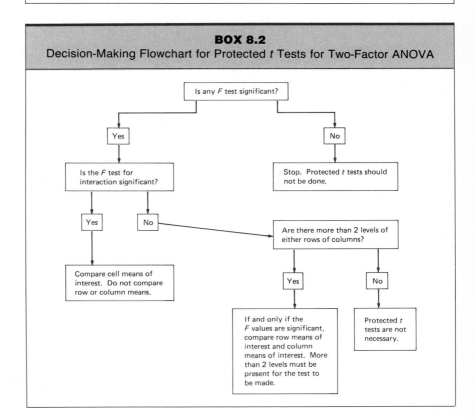

DATA INTERPRETATION AND ANOVA: EXAMPLES

The cell means from a two-factor ANOVA should always be presented in a table or in graphical form. This table or graph will form the basis for interpreting the results of the experiment. The outcome of the ANOVA tests and/or post-hoc tests serve to give credibility to statements that the experimenter makes concerning comparisons within the table or graph.

Let us show how this works by setting up some hypothetical situations involving the graphs in Figure 8.4. Let us suppose that we have performed a two-factor ANOVA on the data represented by panels (a), (b), and (e). Let us suppose that the data in panel (a) came from a 2(gender) × 3(levels of drug) design. Suppose that the F tests for drug and interaction were significant, but the F test for gender was not. What should we say or do? First, the interaction is significant, and, according to the graph, the drug is having the opposite effect for females compared to males. Therefore, we should not attempt to interpret the F's for the main effects (that is, the F's for gender and drug). Instead, we note from the significant interaction that the drug is having the opposite effect for females, as contrasted to males. We should then do post-hoc tests between levels of the drug within each gender condition to see if there is a significant increase for males and a significant decrease for females. Post-hoc comparisons between males and females within drug levels would not be of particular interest.

Now suppose that the results of the experiment turned out as depicted in panel (b), with the F tests for drug and interaction being significant while the F test for gender was not. We again say that the drug is affecting males differently than females, but this time the drug is facilitating female performance, but just not as much as male performance. As before, post-hoc tests should be done within the gender conditions. However, this time the F test for drug has a meaningful interpretation because the direction of the effect is the same for both males and females. Suppose our post-hoc tests showed that both males and females had increased their level of performance as a result of the drug. Then, our concluding statement might be, "The drug facilitated performance for each sex, but the increase was greater for males than for females."

For the data depicted in panel (e), suppose that we have two levels of drug by three levels of age (young, middle, old) design. Suppose further that the F test for interaction was significant, but the F tests for age and drug were not. The significant interaction means that the drug is affecting the age groups differently. However, we cannot tell which age groups are being affected or how until we look at the graph. From the graph, we see that the drug is facilitating the performance of the young and middle-aged, but hindering the performance of the old. We would want to do post-hoc tests comparing drug levels within age groups to see if the young and middle-aged had a significant increase in performance, and if the old had a significant decrease in performance.

As you can see from these examples, data interpretation can get complicated. Additionally, there are often no established guidelines to direct the interpretation. The keys are good judgment and experience. However, there is one element that we haven't mentioned that tends to simplify things. This element is that research is usually motivated by specific hypotheses or interests. Hence, the data analysis is guided by these interests on the part of the researcher.

Finally, a word of advice. As you begin the process of data analysis, use your best judgment and don't be upset if you make some mistakes. We all did.

PSEUDO INTERACTIONS

Occasionally, researchers fall into the trap of inferring that an interaction exists by comparing the results of significance tests for paired comparisons. For example, suppose we give a group of fifth-grade boys and a group of fifth-grade girls a spelling test and find no significant difference in performance. We then test another group of fifth-grade boys and another group of fifth-grade girls using the "gold star incentive system" and find that the statistical test shows that girls do better than boys. Can we conclude that the system worked better for girls than for boys? No, that is a question asking whether the *relative increase* in performance was greater for girls than for boys. That question can only be answered by the test for interaction in a two-factor design. A data example will clarify the major reason for this. Suppose that with no incentive, girls outperform boys by a comparison of means of 30 to 25. This *t* test might have been almost, but not quite, significant. However, with the incentive, girls outperform boys with comparative means of 32 and 26. A *t* test could now show this difference to be significant. However, the relative difference between boys and girls has changed only from 5 (i.e., 30 − 25) to 6 (i.e., 32 − 26), which makes the lines in a graph almost parallel and indicates no interaction. See Figure 8.5. To repeat, one cannot infer that an interaction exists from the results of statistical tests on pairs of cell means; an interaction must be tested within the context of an ANOVA from a factorial design.

Figure 8.5
Illustrating pseudo interaction.

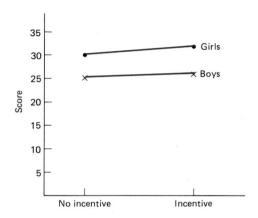

KEY TERMS

factor

interaction

cell

matrix

main effect

error term

effect (in the context of row and
  column means)

predicted cell mean under conditions of no interaction

pseudo interaction

Verbalize the procedure for:

(a)   finding SS_r and SS_{col}
with the definitional formula

(b)   finding row and
column effects

(c)   finding SS_I using the
definitional formula

(d)   doing post-hoc tests;
state all of the rules.

SUMMARY

A two-factor ANOVA combines the conditions of one independent
variable with those of a second independent variable to yield three
separate statistical tests on the data. Two of the tests (on rows and
columns) collapse (i.e., ignore) one factor to test the conditions of the
second factor, as was done with the one-factor ANOVA. The third
test is for an interaction of the two factors, and it is the unique component of the two-factor ANOVA. An *interaction* is defined as the relative effect on a dependent variable coming from different combinations of conditions of two independent variables.

It can be shown that in a two-factor ANOVA that $SS_T = SS_r + SS_{col} + SS_I + SS_W$ (i.e., SS total = SS rows + SS columns + SS interaction + SS error). There is a similar df breakdown: $df_T = df_r + df_{col} + df_I + df_W$. As with the one-factor ANOVA, mean squares (MS) is found by dividing an SS by its corresponding df. MS_W is called the *error term*, and it is the denominator of the F tests.

The key to interpreting the two-factor ANOVA involves the interaction as follows: (a) If the F test for interaction is *not* significant, then the F's for rows and columns can be interpreted as being separate independent tests. (b) If the interaction is significant, row and column tests can be misleading if not interpreted in relation to the interaction. (c) The results of the statistical tests should be interpreted in reference to a table or graph of cell means so that the direction of comparisons (i.e., which is largest) and general patterns become apparent.

To find SS_I by using the definitional formula, one first finds row and column effects (i.e., the grand mean subtracted from row and column means). One then adds the row and column effects that match the row and column containing that cell to the grand mean. When done for each cell, this procedure results in predicted cell means (i.e., $\hat{X}$'s). SS_I is then found by $SS_I = N_{cell} \Sigma(\overline{X} - \hat{X})^2$. If one draws a graph of predicted cell means, the lines within the graph will always be parallel. Therefore, with actual cell means, the extent to which lines in a graph are nonparallel is the extent to which an interaction exists between two factors.

As with the one-factor ANOVA, follow-up tests using the Fisher LSD or protected t test procedures can be performed following a two-factor ANOVA. The key to doing protected t tests following a two-factor ANOVA is whether or not there is a significant interaction. If there is no significant interaction, then tests can be done between row means if the F test for rows is significant. Tests can also be performed between column means if the F test for columns is significant. If the F test for interaction is significant, then comparison between cells can be made. If none of the three F tests in a two-factor ANOVA is significant, then no follow-up paired comparison tests should be performed.

Relative differences between conditions of one factor as a function of the conditions of a second factor can only be tested by the F value for interaction from a two-factor ANOVA. Using the significance and nonsignificance of a series of paired-comparisons to make this kind of inference is not appropriate.

PROBLEMS

(Use $\alpha = .05$ for all statistical tests.)

1. Using the definitional formula, do a two-factor ANOVA on the following data:

	A	B	C
	12	7	7
M	11	9	10
	7	11	7
	14	7	14
F	15	7	12
	13	7	10

2. Use the computational formula to do a two-factor ANOVA on the data in problem 1.

3. An investigator hypothesizes that a certain drug will affect the activity level of hyperactive children in a much different way than for normal children. Activity level was measured by the percentage of time that children were moving some part of their body during a 10-minute reading assignment. The data were as follows:

	Placebo	Drug
	93	33
	97	17
Hyperactive	96	21
	95	30
	94	24
	35	35
	20	52
Normal	15	70
	29	47
	26	71

a   Draw a graph of $\bar{X}$s.

b   Draw a graph of $\hat{X}$s for cells.

c   Based on the graphs, does it appear that an interaction exists? Why?

d   Do a two-factor ANOVA on the data and put the results in a summary table. Include the p column. List the critical F's below the summary table.

e   Do the ANOVA results support the experimenter's hypothesis? Which is the relevant test?

    f   Do the appropriate LSD tests and list the comparison and whether or not $p < .05$ (i.e., $\bar{X}_{NP} > \bar{X}_{HP}$, $p < .05$, etc.).

    g   Describe what happened in the experiment based on the cell means and appropriate tests. By appropriate tests, we mean *F* tests from the ANOVA and post-hoc tests.

4. In studying the ''macho effect,'' an investigator makes three predictions:

    a   Males will tolerate more pain than females in situations where one or more spectators are present.

    b   People in general will tolerate more pain as the number of spectators increases.

    c   The difference in pain tolerance between males and females will increase as the number of spectators increases.

The experimenter does an experiment to test her predictions. Each subject is to place his or her hand in a bucket of ice water for as long as he or she can stand it. Subjects perform the task in front of two, four, or eight spectators. The dependent variable is the number of minutes that a person is able to hold his or her hand in the ice water. The data were as follows:

Spectators

	2	4	8
Males	2.5	2.8	3.5
	2.9	2.9	1.9
	1.6	1.8	2.8
	3.1	2.2	2.7
	2.8	3.3	3.1
	3.0	3.0	3.3
Females	1.9	2.1	1.4
	.6	1.1	1.3
	.7	1.7	1.7
	.9	.6	2.2
	1.7	1.3	3.1
	1.2	1.0	1.5

    a   Draw a graph of $\bar{X}$s.

    b   Draw a graph of $\hat{X}$s for cells.

    c   Based on the graphs, does it appear that an interaction exists? Why?

    d   Do a two-factor ANOVA on the data and put the results in a summary table. List the critical *F*'s below the summary table.

 e Which statistical test from the ANOVA is relevant to which prediction? Which predictions are supported by the results of the statistical tests?

 f Do the appropriate LSD tests and list the comparison and whether or not $p < .05$ (i.e., $\bar{X}_{M2} > \bar{X}_{F2}$, $p < .05$, etc.).

 g Describe what happened in the experiment based on the cell means and appropriate statistical tests.

5. In a task involving the recall of words from a 20-word list, an investigator predicts that imagery-encoding instructions will create better recall than verbal repetition instructions for concrete (e.g., ship) and abstract (e.g., function) words combined. However, he also predicts that the positive effect for imagery instructions will be even greater for concrete words than for abstract words. Using a 2 × 2 factorial design, he obtains the following data:

	Type of Word	
	Concrete	Abstract
Imagery Inst.	14, 15, 15	8, 12, 10
	14, 16, 16	10, 9, 11
Repetition Inst.	10, 12, 8	9, 11, 7
	9, 11, 10	8, 10, 9

 a Draw a graph of $\bar{X}$s.

 b Draw a graph of $\hat{X}$s for cells.

 c Based on the graphs, does it appear that an interaction exists? Why?

 d Do a two-factor ANOVA on the data and put the results in a summary table.

 e Which statistical test from the ANOVA is relevant to which prediction? Which predictions are supported by the results of the statistical tests?

 f Do the appropriate LSD tests and list the comparison and whether or not $p < .05$.

 g Write a paragraph where the results of the F tests are stated in journal format (e.g., The F test for _____ was significant, $F(2,48) = 6.91$, $p < .05$), and also described (e.g., This F test shows that. . . . Further information was obtained from post-hoc tests, which showed that . . .).

6. An investigator believes that attitude change is a function of both the credibility of the speaker and the emotional tone of the speech.

She uses three levels of credibility: low represented by an uniformed student, medium represented by an informed professor, and high represented by an expert in the area. The emotional factor was represented by a passive and a "fiery" presentation. The subject area was the effects of toxic waste. Higher scores reflect greater change toward tough laws concerning disposal. The data reflecting attitude change scores from before and after presentation were as follows:

	Credibility		
	Low	Medium	High
	0	1	2
	0	1	1
Passive	1	0	1
	1	1	2
	0	2	1
	0	0	2
	0	1	4
	1	2	5
Fiery	1	2	4
	0	1	5
	1	2	5
	0	2	5

a Draw a graph of $\bar{X}$s for cells.

b Based on the graph, does it appear that an interaction exists? Why?

c Do a two-factor ANOVA on the data and put the results in a summary table.

d Which statistical test from the ANOVA is relevant to which prediction? Which predictions are supported by the results of the statistical tests?

e Do the appropriate LSD tests and list the comparison and whether or not $p < .05$.

f Describe what happened in the experiment based on the cell means and appropriate statistical tests.

7. In a 2 × 2 matrix, fill in any cell means that meet the following constraints (one matrix per problem):

a zero row effects (i.e., the row means are the same), zero column effects, but nonzero interaction.

b nonzero row effects, zero column effects, nonzero interaction.

c zero row effects, nonzero column effects, zero interaction.

8. Plot graphs of (a), (b), and (c) in problem 7 and note which have parallel lines.

9. Assume that there are 10 scores per call in problem 7. Find SS_r, SS_{col}, and SS_I for (a), (b), (c) in problem 7.

CHAPTER 9

One-Factor Repeated-Measures and Two-Factor Mixed Designs

Important note to student

This chapter is a "hybrid" chapter. It entails an extension of the concepts developed in Chapters 7 and 8. Please review the SS and df breakdown for one- and two-factor ANOVAs as presented in Chapters 7 and 8. This chapter starts with an assumption of this knowledge on the reader's part.

INTRODUCTION

Our first discussion of repeated-measures designs was given under the topic of the dependent groups t test in Chapter 6. Recall that with repeated-measures designs, each subject participates in all of the conditions of an experiment. With respect to the t test, the number of conditions would be two. However, as we saw in the one-factor ANOVA chapter, an experimenter may wish to have an independent variable represented by more than two conditions. We have already seen how a one-factor ANOVA handles the analysis for this type of situation for a between-subjects design (i.e., different subjects participate in each condition of the experiment). We will now explain how within-subject (i.e., repeated-measures) designs are treated with an analysis of variance.

Notice what happened in the analysis of the dependent groups t test. The error term (i.e., $S_{\bar{D}}$) was computed based on each subject's difference score, that is, each subject's score in condition 1 subtracted from his or her score in condition 2. The effect of this procedure was to remove individual differences from the error term. Thus, in a reaction time study, variability coming from the fact that some people are fast while others are slow is not included in the error term (see again Chapter 6). The net result of a smaller error term is to enhance the likelihood of achieving statistical significance. The same principles involving the error term in a matched group t test also apply to a repeated-measures ANOVA. We will demonstrate these principles and the ensuing computations within the context of an example.

Assume, as in Chapter 6, that we are studying the effects of a drug on reaction time. As before, we measure reaction time in hundredths of a second. However, with our current design, we have three conditions (placebo, low, high) involving the quantity of the drug instead of two conditions. In our hypothetical experiment, we have four subjects, each of whom participate in each of the three conditions. We will assume that we have designed the study so that practice or carryover effects from one condition to another are negligible. The data for each of the four subjects from our hypothetical experiment are shown in Table 9.1.

Let us first analyze the data as if the design were between subjects instead of within subjects. From the chapter on one-factor ANOVA we have

$$\text{SS}_T = \Sigma(X - \bar{X}_G)^2 = (27-33)^2 + (33-33)^2 + \cdots + (40-33)^2 = 180$$

$$\text{SS}_B = N_c(\bar{X}_c - \bar{X}_G)^2 = 4[(31-33)^2 + (32-33)^2 + (36-33)^2] = 56$$

$$\text{SS}_W = \Sigma(X - \bar{X}_c)^2 = (27-31)^2 + (33-31)^2 + \cdots + (40-36)^2 = 124$$

$$\text{df}_T = N_T - 1 = 12 - 1 = 11$$

TABLE 9.1
Data from Hypothetical Reaction Time Experiment

	Placebo (P)	Low (L)	High (H)	$\bar{X}$
S_1	27	29	31	29
S_2	33	33	39	35
S_3	29	30	34	31
S_4	35	36	40	37
	$\bar{X}_P = 31$	$\bar{X}_L = 32$	$\bar{X}_H = 36$	$\bar{X}_G = 33$

$$df_B = k - 1 = 3 - 1 = 2$$

$$df_W = k(N_c - 1) = 3(3) = 9$$

$$MS_B = \frac{SS_B}{df_B} = \frac{56}{2} = 28$$

$$MS_W = \frac{SS_W}{df_W} = \frac{124}{9} = 13.78$$

$$F = \frac{28}{13.77} = 2.03$$

Focus now on the numbers within each of the P, L, and H conditions. What makes the scores different within each condition? In other words, what creates the variability within each condition? Part of this variability comes from the fact that some people are relatively fast and others are relatively slow. For example, within the P condition, one person (S_4) had a reaction time of 35 hundredths of a second while another (S_1) had a reaction time of 27 hundredths of a second. This individual difference variability increases the value of SS_W, which in turn will lower the computed F value.

What happens in a repeated-measures design is that individual difference variability is removed from the error term, effectively making the error term smaller, thus increasing the value of the computed F. There are two ways of showing how this occurs. We will demonstrate both. The first technique uses difference scores as with the dependent groups t test. We demonstrate this technique to show that the principle for removing subject variability developed in the dependent groups t test can be extended to the ANOVA. However, there is an equivalent, but simpler, way of removing subject variability from the error term. This simpler technique is demonstrated in Technique 2, and we will base further development using repeated measures (i.e., mixed designs) on this simpler technique.

TECHNIQUE 1: SUBJECT'S DIFFERENCE SCORE TECHNIQUE

First, we convert each subject's scores in each of the three conditions (P, L, H) into difference scores. We do this by subtracting each subject's score in each of the three conditions from the mean of these three conditions. Thus, the difference scores for S_1 are

$$P = 27 - 29 = -2$$

$$L = 29 - 29 = 0$$

$$H = 31 - 29 = 2$$

The difference scores for all four S's would be

	Placebo	Low	High	$\bar{X}$
S_1	-2	0	2	0
S_2	-2	-2	4	0
S_3	-2	-1	3	0
S_4	-2	-1	3	0
	$\bar{X}_p = -2$	$\bar{X}_L = -1$	$\bar{X}_H = 3$	$\bar{X}_G = 0$

These scores reflect each subject's relative performance in each of the three conditions. We now can use these difference scores as the data from our experiment and do a one-factor ANOVA almost as we did before with the between-subjects analysis. The only difference is in the way that df_W is computed. Using these difference scores, we have

$$SS_T = (-2)^2 + (-2)^2 + \cdots + (3-0)^2 = 60$$

$$SS_B = 4[(-2-0)^2 + (-1-0)^2 + (3-0)^2] = 56$$

$$SS_W = [(-2-(-2)]^2 + [(-2-(-2)]^2 + \cdots + (3-3)^2 = 4$$

$$df_T = N_T - 1 = 12 - 1 = 11$$

$$df_B = k - 1 = 3 - 1 = 2$$

$$df_W = (k-1)(N_c - 1) = 2(3) = 6$$

$$MS_B = \frac{56}{2} = 28$$

$$MS_W = \frac{4}{6} = .67$$

$$F = 41.79$$

Note: $df_T \neq df_B + df_W$. **The missing df involves the number of subjects minus 1 (i.e., $S - 1$). Therefore, $df_T = df_B + df_W + df_S$.**

TABLE 9.2

Summary Tables for One-Factor Between-Subjects
and Repeated-Measures Analyses

Between-Subject Analysis

Source of Variation	SS	df	MS	*F*
Between conditions	56	2	28	2.03
Within conditions	124	9	13.78	
Total	180	11		

Repeated-Measures Analysis

Between conditions	56	2	28	41.79
Within conditions	4	6	.67	
Total	180			

Notice from Table 9.2 that SS_B and df_B are the same for the difference score analyses and for the regular score analyses. However, SS_W and df_W are different. Why? The reason that SS_W is smaller for the difference scores is that variability due to individual differences has been removed from the data (i.e., each subject is compared to himself or herself). The reason that df_W is smaller is that the difference scores place additional constraints on the data, namely, that they sum to zero for each subject and sum to zero across subjects. Hence, one df is lost for each subject minus 1, so in our example there are 3 df less for df_W in the difference score data than in the regular data. As you can see from Table 9.1, the removal of individual differences from the error term has created a substantially larger *F* value for the repeated-measures analysis compared with the between-subject analysis. Thus, as we discussed in Chapter 6, a repeated-measures design is more powerful than a between-subject design in situations where a repeated-measures design can be used.

TECHNIQUE 2: SUBTRACTING SS FOR SUBJECTS FROM SS$_W$

Let us now look at an analysis from a slightly different perspective, but one that will lead to the same conclusion. Suppose we find SS for subjects in the same way that we found SS for rows in our two-factor ANOVA chapter. Recall that

$$SS_r = N_{row}(\bar{X}_r - \bar{X}_G)^2$$

Note: **N_{row} is the number of scores per row. If we treat each subject as if it were a row, we have**

$$SS_S = N_{subject}\Sigma(\bar{X}_S - \bar{X}_G)^2$$

Note: $N_{subject}$ **is the number of scores per subject, which equals the number of conditions in which each subject has participated.**

Thus, for the data in Table 9.1,

$$SS_S = 3[(29-33)^2 + (35-33)^2 + (31-33)^2 + (37-33)^2] = 120$$

Observe that $SS_S + SS_W$ (using difference scores) $= SS_W$ (using regular scores); that is, $120 + 4 = 124$. Another way of saying this is that if we remove the individual difference variability (SS_S) from the SS component of the error term of the regular scores (SS_W), we have the SS component of the error term appropriate for our repeated measures analysis. We will label this error term component SS_{WS}. We can put these examples into tree form starting with regular data. The data from our example is in parentheses.

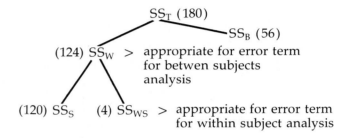

Please note from Chapter 7 that **SS_B** *reflects variability between cell (i.e., condition) means, whereas* **SS_W** *reflects variability of the scores within the cells. Do not confuse this with between- and within-subject variability.*

Degrees of freedom can be broken down in the same way, as shown with df from our example in parentheses.

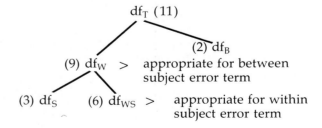

Computational Formula

With respect to computations, the easiest way to compute the sum of squares for the within-subject error term SS_{WS} is to subtract SS_S from SS_W, and the easiest way to compute SS_W is to subtract SS_B from SS_T.

Step 1 Find SS_T:

$$SS_T = \frac{N_T(\Sigma X^2) - (\Sigma X)^2}{N_T}$$

ΣX^2 = sum of all scores squared

$(\Sigma X)^2$ = square of the sum of all scores

Therefore,

$$\Sigma X^2 = 27^2 + 33^2 + \cdots + 37^2 = 13,248$$

$$(\Sigma X)^2 = (27 + 33 + \cdots + 37)^2 = 156,816$$

$$N_T = 12$$

$$SS_T = \frac{12(13,248) - 156,816}{12} = 180$$

If you have a calculator programmed to compute a standard deviation (i.e., σ) then

(a)   Find σ using all scores (12 in our case) as a group.

(b) $SS_T = N_T \sigma^2 = (12)(15) = 180$.

Note: **We found SS from σ, not S, in order to avoid confusing the symbol S with subject as well as standard deviation.**

Step 2 Find SS_B:

$$SS_B = \frac{(\Sigma X_1)^2 + (\Sigma X_2)^2 + \cdots + (\Sigma X_k)^2}{N_c} - \frac{(\Sigma X)^2}{N_T}$$

where X_1 = scores in condition 1
  X_2 = scores in condition 2, etc.
  N_c = number of scores per condition
  k = last group

Therefore,

$$(\Sigma X_1)^2 = (27 + 33 + 29 + 35)^2 = 15,376$$

$$(\Sigma X_2)^2 = (29 + 33 + 30 + 36)^2 = 16,384$$

$$(\Sigma X_3)^2 = (29 + 35 + 31 + 37)^2 = 20,736$$

$$N_c = 4$$

$$N_T = 12$$

So,

$$SS_B = \frac{15,376 + 16,384 + 20,736}{4} - \frac{156,816}{12}$$

$$= 13,124 - 13,068 = 56$$

Alternative method:

(a) Find σ using the condition means as one group.
(b) $SS_B = $ (number of conditions) $N_c\sigma^2 = (3)(4)(4.67) = 56$

> *Step 3* Find SS_W by using the formula $SS_W = SS_T - SS_B$. Thus, in our example, $SS_W = 180 - 56 = 124$. However, a computational check can be obtained by using the following formula:

$$SS_W = \frac{[N_c\Sigma X_1^2 - (\Sigma X_1)^2] + [N_c\Sigma X_2^2 - (\Sigma X_2)^2] + \cdots + [N_c\Sigma X_k^2 - (\Sigma X_k)^2]}{N_c}$$

$$\Sigma X_1^2 = 27^2 + 33^2 + 29^2 + 35^2 = 3884$$

$$\Sigma X_2^2 = 29^2 + 33^2 + 30^2 + 36^2 = 4126$$

$$\Sigma X_3^2 = 31^2 + 39^2 = 34^2 = 40^2 = 5238$$

Thus,

$$SS_W = \frac{4(3884) - 15,376 + 4(4126) - 16,384 + 4(5238) - 20,736}{4}$$

$$= 40 + 30 + 54 = 124$$

> *Step 4* Find SS_S:

$$SS_S = \frac{(\Sigma S_1)^2 + (\Sigma S_2)^2 + \cdots + (\Sigma S_k)^2}{k} - \frac{(\Sigma X)^2}{N_T}$$

where $\Sigma S_1 = $ sum of scores for subject 1
  $\Sigma S_2 = $ sum of scores for subject 2, etc.
   $k = $ the number of conditions

$$SS_S = \frac{(27+29+31)^2 + (33+33+39)^2 + (29+30+34)^2 + (35+36+40)^2}{3}$$

$$- \frac{156,816}{12}$$

$$= 13,188 - 13,068 = 120$$

Alternative method:

(a) Find σ by using subject means (i.e., each subject's mean for the conditions in which they have participated) as a group.

(b) $SS_S = kS\sigma^2$, where k refers to the number of conditions and S refers to the number of subjects ($SS_S = 3(4)(10) = 120$).

Step 5   Find SS_{WS}:

$$SS_{WS} = SS_W - SS_S = 124 - 120 = 4$$

Step 6   Find df_S:

$$df_T = N_T - 1 = 12 - 1 = 11$$
$$df_B = k - 1 = 3 - 1 = 2 \qquad (k = \text{number of conditio}$$
$$df_W = k(N_c - 1) = 3(4 - 1) = 9$$
$$df_S = S - 1 = 4 - 1 = 3 \qquad (S = \text{number of subjects}$$
$$df_{WS} = (k - 1)(N_c - 1) = 2(3) = 6$$

Step 7   Find MS:

$$MS_B = \frac{SS_B}{df_B} = \frac{56}{2} = 28$$

$$MS_{WS} = \frac{SS_{WS}}{df_{WS}} = \frac{4}{6} = .67$$

Step 8   Compute F:

$$F = \frac{28}{.67} = 41.79$$

Step 9   Test H_0:

Null hypotheses for repeated-measures designs are identical to those for between-subjects designs. Thus, for a one-factor design,

$$H_0: \mu_1 = \mu_2 = \cdots = \mu_k$$

For our design, $H_0: \mu_1 = \mu_2 = \mu_3$. The critical F is found by entering the F tables (Table D) at the desired alpha level with df_B across the top and df_{WS} down the side. Thus, the critical $F_{2,6} = 5.14$. Since $41.79 > 5.14$, we reject H_0. The results of this analysis are presented in Table 9.3.

COUNTERBALANCING

As mentioned before, the major disadvantage of repeated-measures designs is practice or carryover effects that may distort the data in some conditions. For example, if a subject is required to solve three different puzzles in three different strategy-induced conditions, the practice coming from early attempts to solve puzzles may affect the attempts to solve the later puzzles. One solution that can reduce the undesirable effect coming from practice or carryover is to counterbal-ance the order in which subjects participate in the conditions of the

TABLE 9.3
Summary Table for Example Experiment Involving Reaction Times as
a Function of Level of Drug

Source of Variance	SS	df	MS	F	p
Total	180	11			
Between cells	56	2	28	41.79	< .05
Within cells	124	9			
Subjects	120	3			
WS error	4	6	.67		

experiment. With counterbalancing procedures, the experimenter first lists all possible orders in which subjects may participate in conditions. The experimenter then assigns equal numbers of subjects to each order. For example, if we have three conditions (e.g., A, B, C), then there are six possible arrangements of orders in which a subject can participate in the three conditions. They are

$$
\begin{array}{ccc}
A & B & C \\
A & C & B \\
B & A & C \\
B & C & A \\
C & A & B \\
C & B & A
\end{array}
$$

If there are N conditions, then there are $N!$ different orders in which these conditions may be presented. With counterbalancing, practice or carryover effects are evenly spread across all conditions. Even when practice or carryover effects are not anticipated, experimenters should counterbalance as a precautionary measure.

POST-HOC TESTS

Protected t tests can be performed in exactly the same manner as was discussed in the chapter concerning one-factor, between-subjects ANOVA. Thus, we have

$$
LSD = t \sqrt{ MS_{WS} \left[\frac{2}{N_c} \right] }
$$

For repeated-measures designs the error term is MS_{WS} instead of MS_W. Therefore, the t value in the LSD formula is found by using a t table with df_{WS} and the desired α level. For our example,

$$
t_{6\,df} (\alpha = .05) = 2.447
$$

$$
MS_{WS} = .67
$$

$$
N_c = 4
$$

Therefore,

$$LSD = 2.447\sqrt{.67(2/4)} = 1.42$$

The pairs of means from our example whose differences exceed the LSD value are placebo versus high and low versus high.

REPEATED MEASURES AND EXPERIMENTAL DESIGN

As you might suspect, two-factor designs and other higher-order designs can also use repeated measures. The principle of removing subject variability from the error term with the resultant increase in statistical power is the same for all orders of experimental design. However, for computational purposes two-factor repeated-measures designs are treated as three-factor designs, with subjects as the third factor. This precludes a computational analysis in a beginning text such as this one. Nevertheless, there is a hybrid design that can be discussed at this time.

Suppose that we expand our interest in the example study on the effects of a drug on reaction time. Suppose that we now ask, "Do levels of a drug affect reaction times differently for males than for females?" This question involves an interest in the interaction between the factors (independent variables) of gender and levels of drug as they affect reaction times. If we have three levels of a drug, the design would be a 2(gender) × 3(levels of drug). Now, can we design this study to take advantage of the increased power of repeated-measures designs? Having each person participate in the six cells created by this design is going to be very difficult, barring some quick sex change surgeries. However, there is a viable alternative. It would be possible for one person to take all of the drug conditions as in our one-factor repeated-measures example. We would need, of course, different people for the gender factor. *Designs that have repeated measures within conditions or levels of one factor and different groups in the conditions or levels of a second factor are called* **mixed designs.**

mixed design:

design containing both a repeated-measures factor and a between-subjects factor

confounded experiment:

an experiment that has results that can be explained by something other than the independent variable manipulation

Mixed designs are very popular in the social sciences because they combine some of the best features of the between-subject designs and the within-subject designs. Subjects in a mixed design participate in fewer of the cells in a two-factor experiment than in a corresponding repeated-measures design, thereby reducing the possibility of practice, carryover effects, or other problems associated with having one subject participate many times in the experiment. Also, the smaller within-subject error term is used for two of the three F tests resulting from a two-factor design, including the F test for interaction.

Another advantage of a mixed design is that it can sometimes turn a *confounded* design into a legitimate one. For example, suppose an investigator predicts that a certain program for teaching math skills will be more effective for gifted children than for average children. if the investigator compares the two groups on test performance following the program, the comparison will be confounded. The children were selected into groups by prior established abilities, and we would expect the gifted group to perform better regardless of the program.

However, within the context of a two-factor mixed design, the confounding disappears. If the investigator administers a pretest, gives the program, and follows the program with a posttest, the investigator would have a 2×2 mixed design. Type of child (gifted or average) would be the between-subject factor. Pre- and posttests would be the within-subject factor. The advantage of this type of design is that the focus is now on the interaction; that is, "Did the gifted group improve more from pre- to posttests than did the average group?" Since the interaction compares the *improvement* in one group to the *improvement* in the second group, the confounding coming from the groups being different at the start is not present.

MIXED DESIGN EXAMPLE

We will illustrate a mixed design by assuming that S_1 and S_2 from our one-way repeated-measures example (Table 9.1) were males and S_3 and S_4 were females. If we analyze these data as a two-factor mixed design, we have the following matrix:

		P	L	H	
M	S_1	27	29	31	$\bar{X}_M = 32$
	S_2	33	33	39	
F	S_3	29	30	34	$\bar{X}_F = 34$
	S_4	35	36	40	

$\bar{X}_P = 31 \quad \bar{X}_L = 32 \quad \bar{X}_H = 36 \quad \bar{X}_G = 33$

Recall from the chapter on two-factor ANOVA that we end up with three tests in a two-factor design: one for rows (i.e., F_r), one for columns (i.e., F_{col}), and one for the interaction (i.e., F_I). We have the same three tests for the two-factor mixed design. Also, the SS breakdown is *identical* to the two-factor between-subjects design with one exception. The variability within cells, SS_W, is broken into two parts, $SS_{S/Gr}$ and SS_{WS}. $SS_{S/Gr}$ (S/Gr is read as subjects within groups) is the variability due to subject differences, and SS_{WS} is the within-cell variability after subject variability has been removed. *Because SS_{WS} is the variability left over in a cell after subject variability has been removed, SS_{WS} is also called* $SS_{residual}$. $SS_{S/Gr}$ is used in the error term in the row or column F test that has different groups of subjects in each of the conditions; that is, $SS_{S/Gr}$ is used in the F test on the between-subject factor. In our example, the between-subject factor is rows, which represent gender. SS_{WS} is used in the error term in the F test that has the same subjects in each of the conditions of that factor. In our example, SS_{WS} would be used in the F test on columns, which tests for differences between levels of the drug. The F test for interaction also uses the within-subject error term since we are comparing relative performance within subjects across conditions.

Since most of the two-factor mixed design analysis duplicates that of the two-factor between-subject analysis, we will elaborate only on the unique aspects of the mixed design analysis. Our analysis consists of two parts. First, we will analyze the data as though both factors were between-subjects factors; that is, we will do a two-factor between-subject ANOVA as described in Chapter 8. Second, we will subtract the variability due to subjects from the SS component of the between-subject error term (SS_W) to get the SS component of the within-subject error term (SS_{WS}).

SS Breakdown

$$SS_T = \Sigma(X - \bar{X}_G) = (27-33)^2 + (33-33)^2 + \cdots + (40-33)^2 = 1800$$

$$SS_r = N_r[\Sigma(\bar{X}_r - \bar{X}_G)^2] = 6[(32-33)^2 + (34-33)^2] = 12$$

$$SS_{col} = N_{col}[\Sigma(\bar{X}_{col} - \bar{X}_G)^2] = 4[(31-33)^2 + (32-33)^2 + (36-33)^2] = 56$$

$$SS_I = N_c\,\Sigma[\bar{X}_c - \hat{X}_c)^2] = 2[(30-30)^2 + (31-31)^2 + \cdots + (37-37)^2]$$
$$= 2(0) = 0$$

Values for $\bar{X}_c$ and $\hat{X}_c$ can be verified from Table 9.4.

TABLE 9.4

		P	L	H	X	Row Effects
M		$\hat{X} = 30$ $\bar{X} = 30$	$\hat{X} = 31$ $\bar{X} = 31$	$\hat{X} = 35$ $\bar{X} = 35$	32	−1
F		$\hat{X} = 32$ $\bar{X} = 32$	$\hat{X} = 33$ $\bar{X} = 33$	$\hat{X} = 37$ $\bar{X} = 37$	34	+1
$\bar{X}$ Column Effects		31 −2	32 −1	36 +3	$33 = \bar{X}_G$	

$$SS_W = \Sigma(X - \bar{X}_c)^2 = (27-30)^2 + (33-30)^2 + \cdots + (40-37)^2 = 112$$

$$SS_{S/Gr} = C_R[\Sigma(\bar{X}_S - \bar{X}_{Gr})^2]$$

where S/Gr represents subjects within groups (i.e., conditions); C_R represents the number of repeated-measures conditions.

Now, $\Sigma(\bar{X}_S - \bar{X}_{Gr})^2$ says to take each subject's mean score and subtract the mean of the group (e.g., the male group or the female group) that contains that subject, square these scores, and then sum the squared scores. Repeating the data with subject means included, we have

		P	L	H	$\bar{X}_S$	$\bar{X}_{Gr}$
M	S_1	27	29	31	29	32
	S_2	33	33	39	35	
F	S_3	29	30	34	31	34
	S_4	35	36	40	37	

$$\bar{X}_G = 33$$

Observe that in our example, males represent one group and females represent a second group.

$$SS_{S/Gr} = 3[(29-32)^2 + (35-32)^2 + (31-34)^2 + (37-34)^2] = 108$$

$$SS_{WS} = SS_W - SS_{S/Gr} = 112 - 108 = 4$$

where SS_{WS} is the SS component of the error term for the within-subject comparisons, and $SS_{S/Gr}$ is the SS component of the error term for the between-subject comparisons. Computational formulas for $SS_{S/Gr}$ and SS_{WS} are given in Box 9.1. A breakdown of SS_T (or df_T) in tree form is

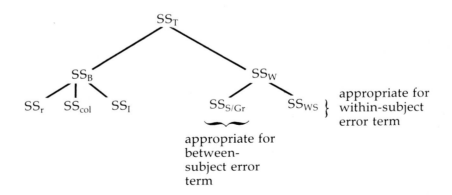

Please note that SS_B and SS_W reflect variability between and within *cells*, not variability between and within *subjects*.

df Breakdown

$$df_T = N_T - 1 = 12 - 1 = 11$$
$$df_r = r - 1 = 2 - 1 = 1$$
$$df_{col} = col - 1 = 3 - 1 = 2$$
$$df_I = (r-1)(col-1) = 1(2) = 2$$
$$df_W = r \times col(N_c - 1) = 6(1) = 6$$
$$df_{S/Gr} = Gr(N_{S/Gr} - 1) = 2(2-1) = 2$$

where Gr is the number of groups in the between-subject variable, and $N_{S/Gr}$ is the number of subjects within a group.

$$df_{WS} = Gr(N_{S/Gr} - 1)(C_R - 1) = 2(1)(2) = 4$$

(or, equivalently, $df_{WS} = df_W - df_{S/Gr} = 6 - 2 = 4$), where C_R represents the number of repeated-measures conditions.

Mean Squares and F's

$$MS_r = \frac{SS_r}{df_r} = \frac{12}{1} = 120$$

$$MS_{col} = \frac{SS}{df_{col}} = \frac{56}{2} = 28$$

$$MS_I = \frac{SS_I}{df_I} = \frac{0}{2} = 0$$

$$MS_{S/Gr} = \frac{SS_{S/Gr}}{df_{S/Gr}} = \frac{108}{2} = 54$$

$$MS_{WS} = \frac{SS_{WS}}{df_{WS}} = \frac{4}{4} = 1$$

$$F_r = \frac{MS_r}{MS_{S/Gr}} = \frac{12}{54} = .22$$

F test for rows: critical $F_{1,2}$ ($\alpha = .05$) = 18.51; .22 < 18.51, therefore retain H_0.

$$F_{col} = \frac{28}{1} = 28$$

F test for columns: critical $F_{2,4} = 6.94$; 28 > 6.94, therefore reject H_0.

$$F_I = \frac{0}{1} = 0$$

F test for interaction: critical $F_{2,4} = 6.94$; 0 < 6.94, therefore retain H_0.

Table 9.5 is a summary table for this mixed design analysis. In mixed design summary tables, the data are organized somewhat differently than shown in the previous tree. Between-subject (not cell) variability is grouped together, and within-subject (not cell) variability is grouped together.

TABLE 9.5
Summary Table for Mixed Design Example

Source of Variation	SS	df	MS	F	p
Total	180	11			
Between subjects	120	3			
rows (gender)	12	1	12	.22	> .05
error (S/Gr)	108	2	54		
Within subjects	60	8			
columns (drug)	56	2	28	28	< .05
interaction	0	2	0	0	> .05
error (WS)	4	4	1		

POST-HOC TESTS

The rationale and procedure for post-hoc tests such as protected t tests for two-factor mixed designs exactly follows those of two-factor between-subjects designs (see Chapter 8). If the F test for an interaction is nonsignificant while the F test for rows is significant, follow-up tests between the row means might be pursued. An identical statement can be made for columns. However, if the F test for an interaction is significant, then individual comparisons between cell means may be desired.

The new element introduced by mixed designs with respect to post-hoc tests is that two error terms are involved instead of one. *When comparisons are between subjects (i.e., the cells or conditions to be compared contain different sets of subjects), the between-subject error term $MS_{S/Gr}$ is used. When comparisons are within subjects (i.e., when the cells or conditions to be compared contain the same set of subjects), the within-subject error term MS_{WS} is used.* Of course, if the situation warrants, two LSD values may be used in mixed designs, one for between-subject comparisons and one for within-subject comparisons. In our example, the LSD appropriate for *between*-subject comparisons between cell means (at $\alpha = .05$) is

$$LSD = t_{2\,df} \sqrt{MS_{Gr} \left[\frac{2}{N_c} \right]} = 4.303 \sqrt{54 \left[\frac{2}{2} \right]} = 31.62$$

The LSD value appropriate for *within-subject* comparisons between cell means (at $\alpha = .05$) is

$$LSD = t_{4\,df} \sqrt{MS_{WS} \left[\frac{2}{N_c} \right]} = 2.776 \sqrt{1 \left[\frac{2}{2} \right]} = 2.78$$

Thus, in comparing the L and H cells for males, we would use the LSD value 2.78. In comparing the H cells for males versus females, we would use the LSD value 31.62. These LSD values for comparing cell means were computed to illustrate the calculations. They would not ordinarily be computed in our example since the interaction test was nonsignificant. However, the F test for columns (i.e., drug level) was significant. Therefore, post-hoc tests on column means could be performed for our example. If this is done, the within-subject error term SS_{WS} should be used in computing the LSD value, since levels of drug is a within-subject variable. An important assumption of these protected t tests is that roughly equivalent Variances exist both within cells and within groups of subjects.

Box 9.1
Summary of Error Term Formulas

One-Factor Repeated Measures

Definitional Formula

$$SS_W = \Sigma(X - \bar{X}_c)^2$$

Take each score, subtract the mean of the condition that the score is in, square, and add for all scores.

Computational Formula

$$SS_W = \frac{N_c \Sigma X_1^2 - (\Sigma X_1)^2 + N_c \Sigma X_2^2 - (\Sigma X_2)^2}{N_c}$$

$$+ \cdots + \frac{N_c \Sigma X_k^2 - (\Sigma X_k)^2}{N_c}$$

where N_c is the number of scores in a condition, $X_1 =$ scores in condition 1, $X_2 =$ scores in condition 2, etc. An equivalent way of finding SS_W is to use the formula $SS_W = SS_T - SS_B$. [For calculators programmed to compute a standard deviation (σ), find σ for the scores in each condition, then convert to SS by using the formula $SS = N_c \sigma^2$. Then $SS_W = SS$ for each condition added together; i.e., $SS_W = \Sigma SS_{condition}$.]

$$SS_S = N_{subject} \Sigma(\bar{X}_S - \bar{X}_G)^2$$

where $N_{subject} =$ the number of scores per subject, $\bar{X}_S =$ the mean of a subject's scores and $\bar{X}_G =$ the grand mean. Thus take each subject's mean minus the grand mean, square, and add for all subjects. Then multiply the total by k.

$$SS_S = \frac{(\Sigma S_1^2) + (\Sigma S_2^2) + \cdots + (\Sigma S_k^2)}{k} - \frac{(\Sigma X)^2}{N_T}$$

Box 9.1 (Continued)

where ΣS_1 = sum of scores for subject 1, ΣS_2 = sum of scores for subject 2, etc., ΣX = sum of all scores, N_T = total number of scores, k = number of conditions. [Alternative: (a) Find σ by using subject means (i.e., each subject's mean for the condition in which they have participated) as a group. (b) $SS_S = kS\sigma^2$, where k refers to the number of conditions and S refers to the number of subjects.]

$$SS_{WS} = \Sigma(D - \bar{D}_c)^2$$

where D = each subject's D score minus the mean D score for that subject. Take each D score and subtract the mean of the D scores for that condition, square, and add for all D scores.

Degrees of freedom
$$df_{WS} = (k - 1)(N_c - 1)$$
where k is the number of conditions and N_c is the number of scores in a condition.

$$SS_{WS} = SS_W - SS_S$$

Mixed Designs

Definitional Formula

Computational Formula

SS_W is computed in identical fashion to SS_W in the one-factor formula (see above) with each cell being considered as a condition in a one-factor design. For example, a 2×3 design would be considered as having six conditions in finding SS_W.

$$SS_{S/Gr} = C_R \Sigma(\bar{X}_S - \bar{X}_{Gr})^2$$

where C_R is the number of repeated measures conditions, $\bar{X}_S$ is the mean for a subject's scores and $\bar{X}_{Gr}$ is the mean of the group that that subject is in.

To find $SS_{S/Gr}$, find $SS_{S/Gr}$ for each group in the design and then add. Thus, $SS_{S/GR} = SS_{S/GR1} + SS_{S/Gr2} + \cdots + SS_{S/Grk}$. The formula for each group is

$$SS_{S/Grj} = \frac{(\Sigma S_1)^2 + (\Sigma S_2)^2 + \cdots + (\Sigma S_N)^2}{k} - \frac{(\Sigma X_{Gr})^2}{N_{Gr}}$$

where Gr_j represents any one group, ΣS_1 = sum of scores for subject 1, ΣS_2 = sum of scores for subject 2, etc., ΣX_{Gr} = the sum of all scores in the group under consideration, N_{Gr} = the total number of scores in the group under consideration. S_N = the last subject in a group. [Alternative: (a) Find SS for each group of subject means by using the formula $SS = N\sigma^2$. (b) Sum the SS values for each group of subject means. (c) Multiply the value in (b) by the number of repeated measures conditions.

$SS_{S/Gr}$ = step c value

$$SS_{WS} = SS_W - SS_{S/Gr}$$

$$SS_{WS} = \Sigma(D - \bar{D}_{cell})^2$$
SS_{WS} is computed as in SS_{WS} for one-factor designs.

BOX 9.2
Summary of SS Symbols

Symbol	Formula	Meaning
SS_T (one or two factor)	$\Sigma(X - \bar{X}_G)^2$ $\bar{X}_G$ = mean of all scores	Variability of all scores when considered as one group
SS_B (one or two factor)	$N_c\Sigma(\bar{X}_c - \bar{X}_G)^2$ N_c = number of scores per cell $\bar{X}_c$ = mean of a cell	Variability due to differences between cell means
SS_W (one or two factor)	$\Sigma(X - \bar{X}_c)^2$	Variability of scores within cells
SS_r (two factor)	$N_r\Sigma(\bar{X}_r - \bar{X}_G)^2$ N_r = number of scores in a row $\bar{X}_r$ = row mean	Variability due to differences between row means
SS_{col} (two factor)	$N_{col}\Sigma(\bar{X}_{col} - \bar{X}_G)^2$ N_{col} = number of scores in a column $\bar{X}_{col}$ = column mean	Variability due to differences in column means
SS_I (two factor)	$N_c\Sigma(\bar{X}_c - \hat{X}_G)^2$ $\hat{X}_c$ = predicted cell mean if no interaction	Variability due to differences between actual cell means and cell means if there were no interaction
SS_S (one factor)	$N_{subject}\Sigma(\bar{X}_S - \bar{X}_G)^2$ $N_{subject}$ = number of scores per subject $\bar{X}_S$ = mean of scores for a subject	Variability due to differences between subject means
SS_{WS}	$SS_W - SS_S$ (one factor)	Variability within cells after subject variability has been removed
SS_{WS}	$SS_W - SS_{S/G}$ (two factors)	
$SS_{S/G}$ (two factors)	$C_R[\Sigma(\bar{X}_S - \bar{X}_{Gr})^2]$ C_R = number of repeated measures conditions $\bar{X}_{Gr}$ = mean of all scores in a group (i.e., a condition of the between-subject factor)	Variability due to differences between subject means within groups

KEY TERMS

error term SS_W
within-subject design SS_{WS}
between-subject design   mixed designs
repeated-measures design SS_W
counterbalancing $SS_{S/Gr}$
   SS_S SS_{WS}

SUMMARY

A repeated-measures design, also called a within-subjects design, is a design wherein each subject participates in every condition of the experiment. The major advantage of a repeated-measures design is that variability due to subject differences is extracted from the error term, thereby increasing the power (i.e., the likelihood of correctly rejecting H_0) of the statistical test. Subject variability (i.e., SS_S) can be computed by treating subject means as row means in a two-factor ANOVA. The appropriate SS for the error term for a repeated-measures design is then found by subtracting SS_S from SS_W (i.e., the SS appropriate for the between-subject error term). Computation of SS_B and MS_B for within-subjects designs is identical to that of between-subjects designs. With repeated-measures designs, counter-balancing procedures are often used. With these procedures, equal numbers of subjects participate in all possible orders by which the conditions of the experiment may be presented. Protected t tests, or tests using LSD can be performed for repeated-measures designs exactly as with between-subjects designs using MS_{WS} as the error term and df_{WS} as degrees of freedom in finding the appropriate critical t value.

Factors in higher-order experimental designs can be either between subjects or within subjects. If designs have at least one between-subjects and one within-subjects factor, they are called *mixed designs*. Thus, a two-factor mixed design has one repeated-measure factor and one between-subjects factor. The computational analysis for the two-factor mixed design is identical to the analysis for the two-factor between-subject design, with one exception: the SS component of the error term for the between-subject (i.e., SS_W) design is broken into two parts. The first part is variability of subjects within groups (i.e., $SS_{S/Gr}$), and this is the SS component of the between-subjects error term. The second part is residual (i.e., left over) variability of SS_W after $SS_{S/Gr}$ has been extracted. This residual variability is called SS_{WS} and is computed by $SS_{WS} = SS_W - SS_{S/Gr}$. SS_{WS} is the SS component of the error term for the within-subjects F tests. Thus, the mixed design has two error terms: one error term, $MS_{S/Gr}$, is appropriate for testing differences between means of the between-subjects factor (i.e., the factor having different groups of subjects in its conditions). The other error term, MS_{WS}, is appropriate for testing differences between means of the within-subjects factor (i.e., the factor having each subject participate in all of its conditions). MS_{WS} is also used as the error term in testing for an interaction.

Protected t tests can be performed following a two-factor mixed ANOVA by using the same rules as for the two-factor between-subjects ANOVA, except that the two error terms require two LSD values. $MS_{S/Gr}$ is used as a component within the LSD for between-subjects comparisons, and MS_{WS} is used as a component for the within-subjects comparisons.

PROBLEMS

(Use $\alpha = .05$ for all statistical tests. Place critical F values below the summary tables.)

1. Do a dependent group t test (Chapter 6) on the following data:

	A	B
S_1	19	24
S_2	23	26
S_3	21	22
S_4	17	28
S_5	20	25

2. Do a within-subject ANOVA on the data in problem 1 and put the results in a summary table. Does $t^2 = F$?

3. An investigator did an experiment where subjects tried to name ink colors of letters that spelled three different types of words: color words different from the ink (symbolized as S for the Stroop effect); numbers (N) ranging from 1 to 9; random words (R). The dependent variable was the time in seconds that it took a subject to read, at top speed and without error, a list of 50 words. Each subject read three lists representing each of the three conditions. The data were as follows:

	S	N	R
S_1	24	18	13
S_2	19	17	12
S_3	20	19	14
S_4	17	18	11
S_5	23	20	15

   a   Do a one-factor repeated-measures ANOVA and put the results in a summary table.

   b   Do the appropriate follow-up paired comparisons tests. Put the means of the three conditions in a table and interpret the results of the tests with reference to the table (e.g., protected t tests on the data in Table X shows $\bar{X}_S > \bar{X}_N$, $p < .05$, etc.).

4. An investigator wished to see if the ability to detect spelling errors would differ as a function of the type of print used in printers. Four print types were used: pica (P), elite (E), bold (B), and dot matrix (D). Each subject participated in all conditions. The error data were as follows:

	P	E	B	D
S_1	7	8	6	9
S_2	5	3	7	2
S_3	3	1	5	5
S_4	0	3	1	1
S_5	4	1	4	5
S_6	1	5	3	7

   a   Make a table of condition means.

   b   Do an ANOVA on the data and put the results in a summary table.

   c   Do follow-up paired comparison tests, if appropriate.

5. An investigator wishes to see if female attractiveness as perceived by males is a function of hair color. She takes copies of a photograph of the same female face and colors the hair either red, blonde, brown, or black. These four photographs are then randomly interspersed with 50 other photographs of women. Seven college age males are randomly chosen and asked to rate the attractiveness of the four target faces and the 50 fillers on a 9-point scale, with 9 being the most attractive. The data for the four target faces were as follows:

	Red	Blonde	Brown	Black
S_1	9	8	7	7
S_2	7	6	6	7
S_3	6	6	6	7
S_4	5	8	6	6
S_5	8	8	7	7
S_6	7	7	9	9
S_7	7	7	7	7

a   Make a table of condition means.

b   Do an ANOVA on the data and put the results in a summary table.

c   Do follow-up paired comparison tests, if appropriate.

6. An investigator predicts that emotionally charged words will be better retained in memory over long periods of time than will neutral words. He gives one group of subjects 30 emotional words at a 10-second rate and measures recall twice, once immediately following the list presentation and then two weeks later. A different group follows the same procedure using 30 neutral words. The data were as follows:

		Immediate Recall	Two-Week Recall
	S_1	17	16
Emotional	S_2	15	17
words	S_3	11	8
	S_4	13	12
	S_5	15	11
Neutral	S_6	12	8
words	S_7	16	9
	S_8	18	12

a   Draw a graph of $\bar{X}_S$ for cells.

b   Based on the graph, does it appear that an interaction exists? Why?

c   Do a two-factor mixed ANOVA on the data and put the results in a summary table.

d   Do the ANOVA results support the experimenter's hypothesis? Which is the relevant test?

e   Do the appropriate post-hoc tests and list the comparison and whether or not $p < .05$ (e.g., $\bar{X}_{EI} > \bar{X}_{EW}$, $p < .05$, etc.).

f   Describe what happened in the experiment based on the cell means and statistical tests.

7. On the basis of an imagery test, an investigator divides people into two groups: good imagers and poor imagers. She then hypothesizes that (a) the good imagers will be better able to solve three-dimensional puzzles than will the poor imagers; (b) the good imagers will show more improvement over a series of puzzles. The following represent minutes to solution for four puzzles.

		1st Puzzle	2nd Puzzle	3rd Puzzle	4th Puzzle
	S_1	2.1	2.0	1.6	1.4
	S_2	1.8	1.9	1.6	1.3
Good	S_3	2.5	2.2	1.8	1.5
imagers	S_4	3.3	3.1	2.6	2.0
	S_5	1.7	1.3	1.1	.8
	S_6	3.2	3.0	3.3	3.1
	S_7	2.8	2.6	2.5	2.4
Poor	S_8	3.5	3.4	3.2	3.2
imagers	S_9	3.0	3.2	2.6	2.6
	S_{10}	3.7	3.3	3.6	3.3

a Draw a graph of $\overline{X}_S$ for cells.

b Based on the graph, does it appear that an interaction exists? Why?

c Do a two-factor mixed ANOVA on the data and put the results in a summary table.

d Which statistical test from the ANOVA is relevant to which prediction? Which predictions are supported by the results of the statistical tests?

e Do the appropriate LSD tests and list the comparison and whether or not $p < .05$.

f Describe what happened in the experiment based on the cell means and statistical tests.

8. An investigator predicts that instructions to make images of words will be a relatively bigger help in the recall of those words than in the recognition of those words when compared to the same words given under instructions to repeat the words aloud. Forty words were presented to subjects under instructions to either form images of the words (i.e., if the word were *ship,* make an image in your mind of a *ship*) or to repeat the words aloud. The words were presented at a 10-second rate. A recall test was then given (e.g., write down all of the words that you can remember) followed by a recognition test (e.g., of these 80 words, circle the 40 that were in the list). The data were as follows:

		Recall	Recognition
	S_1	13	36
	S_2	18	37
Imagery	S_3	16	37
	S_4	17	38
	S_5	10	36
	S_6	9	34
Repeat	S_7	11	33
	S_8	10	37

a   Draw a graph of $\overline{X}_S$ for cells.

b   Based on the graph, does it appear that an interaction exists? Why?

c   Do a two-factor mixed ANOVA on the data and put the results in a summary table.

d   Do the ANOVA results support the experimenter's hypothesis? Which is the relevant test?

e   Do the appropriate LSD tests and list the comparison and whether or not $p < .05$.

f   Describe what happened in the experiment based on the cell means and statistical tests.

9. Under what circumstances would the statistical advantage of doing a repeated-measures ANOVA as opposed to a between-subjects ANOVA be negligible?

CHAPTER 10

Correlation

CORRELATION VERSUS HYPOTHESIS TESTING

Let us examine two types of questions. First, does practice affect the time that it takes a person to solve five-letter anagrams? Second, how strongly related to five-letter-anagram solution time is practice? The first question involves hypothesis testing that we discussed in earlier chapters. We could answer that type of question with an experimental design wherein we have data taken from a group of subjects during the initial stages of practice and then again during a later stage of practice. We could then use a dependent group's t test or a repeated-measures ANOVA to see whether a statistically significant difference existed between the two groups. If H_0 were rejected and if we have a solid experimental design, we could state with confidence (95%, if the chosen α level were .05) that practice improved (assuming times for the high practice groups were lower than for the low practice groups) performance.

However, tests of hypotheses say nothing about how strongly related the independent variable is to the dependent variable. Thus, in our example, rejecting H_0 does not imply that practice helped a lot. It only implies that practice had some nonzero effect. To answer questions about the strength of relationships, we need to use a different technique. Such a technique is called *correlation*. *The term **correlation** means relationship between variables. A **correlation coefficient** is a numerical value ranging from -1 to $+1$ that is found by using a particular mathematical technique that allows us to talk about the magnitude of relationships.* The particular type of correlational technique that we will study in this chapter results in what is called the *Pearson correlation coefficient.* It is based on the linear relationship between two variables.

Pearson correlation:

a correlational technique whereby squared distances from points in a scatterplot to a straight (i.e., regression) line are minimized

Before embarking on an explanation of this technique, we note one further difference between correlational methods and hypothesis testing procedures. With hypothesis testing, we had a clearly defined independent variable and a clearly defined dependent variable. ***With correlational techniques, independent and dependent variables are often arbitrary.*** With our example on the effects of practice on puzzle-solving time, the independent variable is clearly practice and the dependent variable is clearly puzzle-solving time. However, we could also talk about the relationship between grade point average (GPA) and a measure of athletic ability or the relationship between the cost of people's homes and the cost of their cars. In these latter examples, neither of the variables being related could be thought of as being manipulated or causing something to happen with respect to the other. Thus, they are simply variables being related without independent and dependent variable tags. Therefore, two different purposes of experimental techniques emerge. The purpose of experimental designs that use hypothesis testing procedures is to attribute causality to the independent variable or variables. The purpose of correlational techniques is to determine the strength of relationship between variables. However, it is possible for an experimental technique to use both procedures, as discussed in Chapter 12.

CORRELATION AND CAUSATION

Correlational procedures do not allow causality to be attributed to the variables being correlated. If we attribute causality to one of the variables being related, we do so on a basis other than the correlational technique itself. However, correlational techniques are not unique with respect to not allowing the determination of causality. *No statistical test by itself can be used to determine causality.* When we reject

H_0 with a hypothesis testing procedure, we are in effect saying that some nonchance event has occurred. With a "tight" experimental design, we infer that the nonchance event was caused by the independent variable. The problem with correlational techniques is that often the situation involving the correlation is such that variables other than the ones being correlated are influencing one or both of the correlated variables. For example, suppose that we find a strong relationship (i.e., correlation) between the number of hours that children watch violent TV programs and their aggressive acts during play situations. We cannot say for sure that watching violent TV shows is causing aggressive play or that aggressive play is causing the watching of violent TV shows. It could be that some children have aggressive personalities for reasons having nothing to do with TV shows or play, and that children with this type of personality like to watch violent TV shows and play aggressively. Thus, attributing causality to one of the correlated variables is not possible.

SCATTERPLOTS

bivariate:

something that has two variables associated with it

case:

something that has variables associated with it (e.g., a point in a scatterplot)

*Correlation of the type that we are studying is said to be **bivariate**. By this, we mean that each case (person, for example) has two variables associated with it.* Thus, we can calculate a correlation coefficient between study time and GPA provided that we can get a group of *cases* (students in this example), and from each case get a measure of the two variables of interest. We could not talk about (or find) a correlation if we get study time measures from 10 students and GPAs from a different 10 students. Since correlational data are bivariate, they can be plotted on a graph having two axes, each axis representing values of one of the variables being correlated. This type of graph is called a *scatterplot*. Each case will represent a point (i.e., dot) on the graph. Figure 10.1 shows five scatterplots, each using 10 hypothetical cases, relating study time with GPA. Look at each scatterplot and choose the one that you think would most likely represent real data from 10 randomly selected college students.

In discussing this question, let us first describe what is happening within each graph. Within a scatterplot, strength is determined by the extent to which the points within the graph form a straight line. Graph *a* represents every student's fantasy: A very strong relationship exists between study time and GPA such that the less you study, the higher is your GPA. Notice also that a strong relationship enables good prediction. In graphs *a* and *e*, knowing how many hours a day that a person studies, will enable one to make a precise prediction concerning that person's GPA. Now, contrast graphs *b* and *d* with *a* and *e*. In *b* a relationship between study time and GPA still exists, but

Figure 10.1
Hypothetical scatterplots.

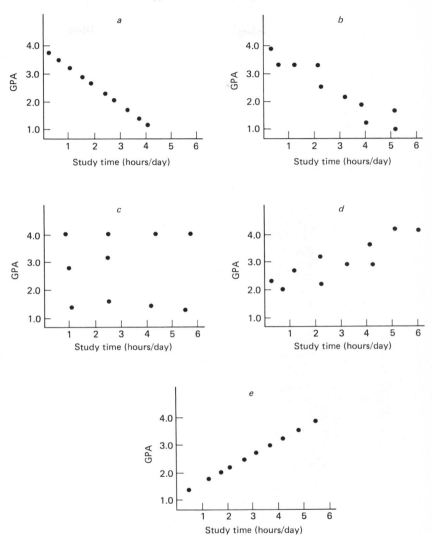

not so strongly. The points do not fall in a straight line and if we know the study time for an unknown person, we cannot say exactly what the GPA would be for that person. However, generally good grades go with less study. The reverse is true for graph *d*, where generally the more study, the better the GPA. Look now at graph *c*. Here, there seems to be little or no relationship between hours of study and GPA. For the four people having the highest GPAs, one studied a lot, one studied little, and two studied in between. The same thing was true for the four people having the lowest GPAs; one studied a lot, one studied little, with two studying in between.

Notice that the data in graphs *a* and *b* "point" in the opposite direction to that of graphs *d* and *e*. If one draws a straight line through the center of a series of points, and this line has *negative slope* (i.e., points left), the relationship between the two variables is said to be *negative*. If the slope of the line is to the right, the relationship is said to be *positive*. The mirror image of a scatterplot of any positive relationship will be a negative relationship of the same magnitude.

WHERE DOES THE SCATTER WITHIN A SCATTERPLOT COME FROM?

Let us now examine our original question: Which scatterplot would most likely represent real data? Based on your personal experience, you would probably say something like: "the more that I study, the better are my grades." Therefore, you might decide to choose *e*. However, even if your reasoning is correct, the correct answer is *d*, not *e*. *One reason is because of individual differences between people (i.e., subject variability).* Jim may need much less study time to get the same grades as George. Individual differences show up in scatterplots by making the scatter (i.e., variability of points from the center line) of points larger. *The other major contributor to the scatter of points is varying factors within an organism.* In my example, you may study long, but have a poor background in the material. Or perhaps you were ill or overly anxious on exam days. These and many other reasons create variability in one person's performance that will create scatter or variability. For all of these reasons, one rarely finds scatterplots approaching straight lines, using real data with numerous cases.

INTERPRETING DIRECTIONALITY

A positive relationship between variables means that as one variable gets larger, corresponding values of the second variable also get larger, as in graphs *d* and *e*. A negative relationship between variables means that as values of one variable get larger, corresponding values of the second variable get smaller, as in graphs *a* and *b*. In many situations, whether a relationship is positive or negative is arbitrary and, in either case, the interpretation is the same. For example, suppose that we wish to look at the relationship between study time and scores on a spelling test. We will assume that the relationship is such that the more one studies, the better is the score. However, spelling scores can be presented in terms of number correct or in terms of the number of errors. Figure 10.2 shows scatterplots of the same data with panel *a* representing scores in terms of number correct (70 points maximum) and panel *b* representing the same scores in terms of number

Figure 10.2

Scatterplots of the same date measured in terms of number correct (*a*)
and number of errors (*b*).

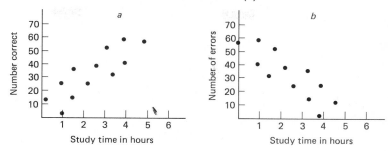

of errors. Notice that *a* represents a positive relationship and *b*
represents a negative relationship, but the interpretation of both scatterplots is the same, namely, that practice is positively related to performance.

PEARSON CORRELATION COEFFICIENTS

Scatterplots are a convenient way to visualize the general relationship
between two variables, but they do not allow a precise numerical
statement to be made about the size of the relationship. By contrast, a
Pearson correlation coefficient utilizes a mathematical procedure that
assigns values that range from -1 to $+1$ to the strength of the correlational relationship: -1 represents a perfect negative relationship, 0
represents no relationship, and $+1$ represents a perfect positive relationship. All values between $+1$ and -1 are possible, with the larger
the absolute value (i.e., ignoring the sign) of the number, the stronger
the relationship between the variables.

The rationale for the development of the Pearson correlation
coefficient is to first construct a straight line through the middle of a
scatterplot such that the squared distances between each point and
the line (called a *regression line*) is minimized. It will suffice for purposes of conceptual understanding to think of a regression line as simply a best-fit straight line through the scatterplot points. The squared
distances between the points and the regression line are mathematically related in such a way that the *smaller* the distances (i.e., less variability), the larger the correlation coefficient up to a maximum of $+1$.
Because it is the distances from points in a scatterplot to a *straight* line
(as opposed to some curved line) that are minimized, the type of
correlation that we study in this chapter is called a *linear* correlation.
Linear correlations are so common that if one simply sees *correlation*,
one can safely assume that linear correlation is implied. Curvilinear
exceptions are always noted.

regression line:

(also called prediction
line) best-fit straight
line through a scatter-
plot used for predicting
values of one variable
based on values of the
other variable

The mathematical formula for finding the Pearson correlation r is

$$r = \frac{\Sigma Z_x Z_y}{N} \qquad\qquad 10.1$$

Equation 10.1 says that r equals the sum of the cross products of the standardized (see Chapter 5) scores of each variable divided by the number of pairs of scores. Your understanding of r will be greatly enhanced if you pay close attention to the following examples. A teacher wants to find the correlation between homework scores (we will call this the X variable) and quiz scores (the Y variable) for the six students in her class. We will draw the scatterplots and compute the correlation coefficients for the following hypothetical situations.

Situation a

	X	Y
S_1	19	15
S_2	18	14
S_3	17	13
S_4	15	11
S_5	14	10
S_6	13	9

Let us now look at these data. Do high scores on X relate to high scores on Y, and low scores on X to low scores on Y? Next, how strong is this relationship? Notice that each X score is 4 units larger than the corresponding Y score. Now look at the data within the scatterplot in Figure 10.3.

Figure 10.3
Scatterplot showing quiz-homework relationship.

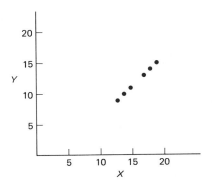

The points in Figure 10.3 form a straight line sloped right, so we would expect an r of $+1$. To compute r, we need to convert each "raw" score for X and Y into a Z score. We then multiply the corresponding Z scores, add these Z scores together and divide by N. Let's do it.

First, we find the standard deviation (σ) for X and Y.

X	$(X-\bar{X})$	$(X-\bar{X})^2$	Y	$(Y-\bar{Y})$	$(Y-\bar{Y})^2$
19	$+3$	9	15	$+3$	9
18	$+2$	4	14	$+2$	4
17	$+1$	1	13	$+1$	1
15	-1	1	11	-1	1
14	-2	4	10	-2	4
13	-3	9	9	-3	9
$\bar{X} = 16$			$\bar{Y} = 12$		

$$\Sigma(X-\bar{X})^2 = 28 \qquad \Sigma(Y-\bar{Y})^2 = 28$$
$$N = 6 \qquad\qquad N = 6$$
$$\sigma_X = \sqrt{\frac{28}{6}} = 2.16 \qquad \sigma_Y = \sqrt{\frac{28}{6}} = 2.16$$

Notice that the σ's for X and Y are the same because identical score spreads are involved in each group. This will not ordinarily be the case.

Using the Z score formula (Chapter 5) $Z = (X - \bar{X})/\sigma$, we convert each raw score into a Z score. For example, the raw score, 18, yields a Z score of

$$Z = \frac{18 - 16}{2.16} = \frac{2}{2.16}$$

Converting all of the X scores and all of the Y scores into Z scores and multiplying the two scores together on a subject-by-subject basis, we have

	Z_X	Z_Y	$Z_X Z_Y$
S_1	$\dfrac{+3}{2.16}$	$\dfrac{+3}{2.16}$	$\dfrac{9}{4.67}$
S_2	$\dfrac{+2}{2.16}$	$\dfrac{+2}{2.16}$	$\dfrac{4}{4.67}$
S_3	$\dfrac{+1}{2.16}$	$\dfrac{+1}{2.16}$	$\dfrac{1}{4.67}$
S_4	$\dfrac{-1}{2.16}$	$\dfrac{-1}{2.16}$	$\dfrac{1}{4.67}$
S_5	$\dfrac{-2}{2.16}$	$\dfrac{-2}{2.16}$	$\dfrac{4}{4.67}$
S_6	$\dfrac{-3}{2.16}$	$\dfrac{-3}{2.16}$	$\dfrac{9}{4.67}$

$$\Sigma Z_X Z_Y = \frac{9+4+1+1+4+9}{4.67} = \frac{28}{4.67}$$

$$r = \frac{\Sigma Z_X Z_Y}{N} = \frac{28/4.67}{6} = \frac{28}{28} = 1$$

Notice that the key to $r = 1$ is that the numerator of $\Sigma Z_X Z_Y$ equals 28. This happened because (a) all of the numbers being added together were positive, and (b) this arrangement of X and Y numbers maximizes the value of the sum of the cross products. The latter is not intuitively obvious, but notice what happens if all scores remain the same, with the exception of the person with the highest X score now having the second highest Y score, and vice versa as in Situation b.

Situation b   We leave the X scores the same but interchange position 1 and 2 of the Y scores. The data now look as follows:

	X	Y
S_1	19	14
S_2	18	15
S_3	17	13
S_4	15	11
S_5	14	10
S_6	13	9

We now put the data into a scatterplot (Figure 10.4). Notice that the points in Figure 10.4 almost, but not quite, form a straight line. Since the numbers within each X and Y set are the same as before, just slightly rearranged, the σ for both groups will remain the same. Therefore, let us look now at the Z scores and their cross products:

Figure 10.4

Scatterplot for Situation b.

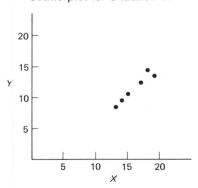

	Z_X	Z_Y	$Z_X Z_Y$
S_1	$\dfrac{3}{2.16}$	$\dfrac{+2}{2.16}$	$\dfrac{6}{4.67}$
S_2	$\dfrac{+2}{2.16}$	$\dfrac{+3}{2.16}$	$\dfrac{6}{4.67}$
S_3	$\dfrac{+1}{2.16}$	$\dfrac{+1}{2.16}$	$\dfrac{1}{4.67}$
S_4	$\dfrac{-1}{2.16}$	$\dfrac{-1}{2.16}$	$\dfrac{1}{4.67}$
S_5	$\dfrac{-2}{2.16}$	$\dfrac{-2}{2.16}$	$\dfrac{4}{4.67}$
S_6	$\dfrac{-3}{2.16}$	$\dfrac{-3}{2.16}$	$\dfrac{9}{4.67}$

$$\Sigma Z_X Z_Y = \frac{6+6+1+1+4+9}{4.67} = \frac{27}{4.67}$$

$$r = \frac{27/4.67}{6} = \frac{27}{28} = .96$$

The numerator of $\Sigma Z_X Z_Y$ is now 27 instead of 28, therefore $r < 1$. If you play with the numerators of the Z score cross products for X and Y, you will find no arrangement that will sum to more than 28. Now, notice what happens if we invert all of the Y scores from Situation a, where we had a cross product numerator sum of 28. The person with the highest X score now has the lowest Y score. The person with the second highest X score has the second lowest Y score, and so on. All of the signs in the Z_Y column change, but the numbers all remain the same. We therefore end up with

$$\Sigma Z_X Z_Y = \frac{-9-6-4-1-1-4-6-9}{4.67} = \frac{-28}{4.67}$$

$$r = \frac{-28/4.67}{6} = \frac{-28}{28} = -1$$

The r is now -1 instead of $+1$.

Situation c Suppose we scramble the Y values so that there is little consistency between the sizes of the Y values and the corresponding X values. Suppose the situation were now as follows:

	X	Y	Z_X	Z_Y	$Z_X Z_Y$
S_1	19	14	$\dfrac{3}{2.16}$	$\dfrac{2}{2.16}$	$\dfrac{6}{4.67}$
S_2	18	9	$\dfrac{2}{2.16}$	$\dfrac{-3}{2.16}$	$\dfrac{-6}{4.67}$
S_3	17	11	$\dfrac{1}{2.16}$	$\dfrac{-1}{2.16}$	$\dfrac{-1}{4.67}$
S_4	15	15	$\dfrac{-1}{2.16}$	$\dfrac{3}{2.16}$	$\dfrac{-3}{4.67}$

$$S_5 \quad 14 \quad 10 \quad \frac{-2}{2.16} \quad \frac{-2}{2.16} \quad \frac{+4}{4.67}$$

$$S_6 \quad 13 \quad 13 \quad \frac{-3}{2.16} \quad \frac{1}{2.16} \quad \frac{-3}{4.67}$$

$$\Sigma Z_x Z_Y = \frac{6 - 6 - 1 - 3 + 4 - 3}{4.67} = \frac{-3}{4.67}$$

$$r = \frac{-3/4.67}{6} = \frac{-3}{28} = -.11$$

We have a very small r, $-.11$. The scatterplot of these data is shown in Figure 10.5. As you can see, there is not an obvious linear pattern to these data.

Figure 10.5
Scatterplot for Situation c.

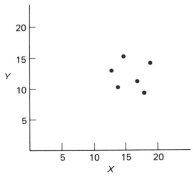

PROPERTIES OF r

What would happen to r if, in any of the three situations we have described, we subtract 5 from each X score and add 73 to each Y score? As we saw in Chapter 5, we can add or subtract a constant (i.e., any single number) from a set of scores without changing the σ or deviations of scores from the mean. Thus, the respective Z scores for X and Y would remain the same as would r. Thus: *Adding or subtracting a constant from X or Y does not change r.* Note, also, that the X scores and the Y scores are standardized (i.e., converted into Z scores) before the r is computed. Thus, the X and Y variables are converted into the same relative scale before r is computed. Therefore, a meaningful r can be computed even if the X and Y raw score scales are very different. For example, a meaningful r could be computed between quiz scores ranging from 1 to 20 and SAT scores ranging from 300 to 700.

COMPUTATIONAL FORMULA FOR r

As we have seen many times, formulas to develop concepts are not always the easiest to work with when using ordinary calculators. It can be shown that an equivalent formula for r is

$$r = \frac{N\Sigma XY - (\Sigma X)(\Sigma Y)}{\sqrt{[N\Sigma X^2 - (\Sigma X)^2][N\Sigma Y^2 - (\Sigma Y)^2]}} \qquad 10.2$$

Equation 10.2 is often called the *raw score formula* because it can be used with the raw scores without converting them into Z scores. We will illustrate the raw score formula by using it to again find r for Situation b. Repeating the data and showing the relevant calculations, we have

X	Y	XY	X²	Y²
19	14	266	361	196
18	15	270	324	225
17	13	221	289	169
15	11	165	225	121
14	10	140	196	100
13	9	117	169	81

$$\Sigma X = 96 \quad \Sigma Y = 72 \quad \Sigma XY = 1179 \quad \Sigma X^2 = 1564 \quad \Sigma Y^2 = 892 \quad N = 6$$

Note. N **is the number of pairs of scores.**

$$r = \frac{6(1179) - (96)(72)}{\sqrt{[6(1564) - (96)^2][6(892) - (72)^2]}}$$

Simplifying, we have

$$r = \frac{7074 - 6912}{\sqrt{[168][168]}} = \frac{162}{168} = .96$$

This value for r is the same as the one we obtained by using Equation 10.1.

INTERPRETING r

The Magnitude of r

Pearson correlation:

a correlational technique whereby squared distances from points in a scatterplot to a straight (i.e., regression) line are minimized

As we have already seen, the larger the r, the stronger is the relationship between two variables. However, values of r do not form a ratio scale. We cannot say that an r of .50 is twice as strong as an r of .25. We can only say that an r of .50 shows a stronger relationship between two variables than an r of .25. However, there is a way to convert r into a ratio scale with respect to strength of relationship. We do this simply by squaring r. This measure, r^2, is called the *percent of Variance explained for two correlated variables.* Thus, an r of .8 explains 64% of the variance, and an r of .4 explains 16%, or one fourth the variance of the .8 correlation (16 = 64/4). The relationship between the size of r and the magnitude of percent of variance explained is shown in Figure 10.6.

Figure 10.6
Relationship between size of *r* and percent Variance explained.

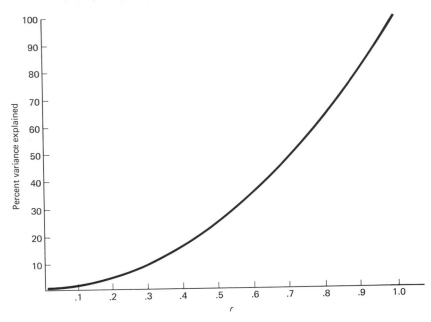

Observe that the curve in Figure 10.6 accelerates upward after appearing flat because of very low initial values. Correlations below .2 explain only a tiny proportion of variance. Notice how for each unit that the correlation coefficient increases, there is a relatively larger increase in the percentage of Variance explained. (This is why the curve accelerates upward.)

So what does percent of Variance explained mean? It means the extent to which knowing values of one variable will allow one to account for score spread (i.e., variability) of the second variable. For example, using 50 cases, suppose we find an *r* of .7 between hours of practice and performance on a motor skills task. If we simply look at these 50 motor skill scores, we will see that there is likely a large amount of variability (i.e., some high scores, some low, and some in the middle). Why are the scores not all the same? They are not the same because people are different, because of faulty testing, because of fatigue factors, and so on. But the scores are also different because of differences in practice. An *r* of .7 means that r^2 is .49, and an r^2 of .49 in our example means that 49% of the differences that we see (measured in terms of Variance) in the motor skills task is associated with practice. The other 51% is associated with any or all of the other variables.

Restricting the Range of *r*

The magnitude of the correlation coefficient depends on the ranges of the *X* and *Y* variables. If either or both of the ranges are restricted, the *r* will be lowered. For example, it is well known that Graduate Record Exam (GRE) scores do not correlate highly with grades of graduate students at highly rated universities. This seems counterintuitive until we look more closely at the situation. Suppose that every college graduate took the GRE and was also admitted to a top graduate school. We might very well find a strong relationship between combined verbal and quantitative GRE scores and grades in graduate school. A hypothetical scatterplot might look as in Figure 10.7.

Suppose, however, that top graduate schools have an admission cutoff of 1300 for GRE scores. Now, the range of GRE scores is severely restricted. The box within the scatterplot indicates how the restricted scatterplot looks. Notice that within the box there is no strong linear relationship, but points are scattered everywhere. Further, if you were to compute an *r* for just the points within the box, you would find a very small correlation coefficient. Thus, finding a low correlation between GRE scores and graduate school grades is not surprising.

A correlation coefficient can also be increased by omitting the scores in the middle range of one variable and computing *r* based on scores at both ends. For example, if we omit the points in our scatterplot between say 900 to 1100, the *r* would be higher than the *r* that includes all of the points. Since the magnitude of *r* can be influenced

Figure 10.7

Restriction of range effects on *r*.

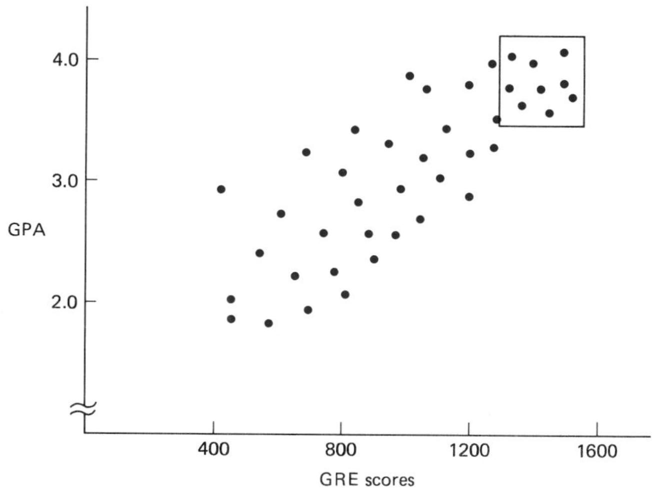

in either direction by the range of the variables being correlated, a complete interpretation of r requires knowledge of the range of both variables.

Sample size

Suppose that we wish to find the correlation between the number of hours that children spend watching violent TV shows and the amount of aggression exhibited in play situations. We will assume that we have a reliable way to measure both the number of hours of violent TV watching and aggression during play. We will obtain a measure from each child for each variable and compute a correlation coefficient. The question now is, ''how many children do we need to give an adequate estimation of the true correlation?'' Everything that we have said about estimating population parameters from sample statistics applies in this situation. We would compute a correlation coefficient based on a random (hopefully) sample of children. We would use that correlation to estimate the correlation for children in general. How large should this sample be to get a reliable estimate? Just as in estimating population means from sample means, the answer is the larger the better. As we discussed earlier, the scatter (i.e., variability) in a scatterplot comes about for a variety of reasons. A correlation coefficient based on only a few cases can vary greatly because of random influences having nothing to do with the two variables being related. How many cases is enough to ensure that a sample r is representative of a population r is somewhat arbitrary, but a sample size of 30 or more cases is usually considered desirable.

Testing r for significance

Imagine the following scenario. There is really no relationship between watching violent TV shows and aggression in children. There are children who watch a lot of violent TV and are aggressive, and children who watch a lot of violent TV and are not aggressive. Likewise, some children watch little violent TV and are aggressive, whereas other children watch little violent TV and are not aggressive. Suppose that we randomly select a sample size N (say 10) from this population of children and compute r between violent TV watching and aggression. We would expect that this r would be close to, but not necessarily, zero, because by chance we may have had an over-representation of, say, high violent TV watchers who are aggressive and low violent TV watchers who are not aggressive. This selection would yield a large positive correlation. In other words, correlations based on samples from a population will vary. Therefore, if we repeated this procedure (i.e., drawing 10 cases from a population having zero correlation) several thousand times, we would have a frequency distribution of sample r's. From this frequency distribution, we could tell how large r has to be before we would conclude that an r of that magnitude is a rare event (i.e., occurs less than 5% of the time using $\alpha = .05$). Thus, we could find a cutoff r from which we could compare an r based on a sample of size N (10 in our example) to see if this was an unlikely r to have occurred if the population correlation was zero.

Happily, there is an easier way to find cutoffs. It turns out that the t distributions have approximately the same shape as distributions of r's, based on large numbers of samples of r's of size N from a population where r was zero. With an appropriate conversion, one can transform a sample r into a computed t score, which we can then compare to a critical t (based on a chosen α level) to determine if the sample r is significantly different from zero. The formula for transforming r into a t value is

$$t = r \frac{\sqrt{N-2}}{\sqrt{1-r^2}} \qquad \qquad 10.3$$

Example Suppose we randomly select 30 five-year-old children and correlate the time spent watching violent TV with a measure of aggression on the playground. We find an r of .30. Is this r significantly different from zero (i.e., H_0: population $r = 0$) at $\alpha = .05$ with a nondirectional test?

Using an r of .3 in Equation 10.3, we have

$$t = \frac{.3\sqrt{30-2}}{\sqrt{1-.3^2}} = 1.66$$

The critical t for $N - 2$ (i.e., $30 - 2$) df with $\alpha = .05$ and a nondirectional test is 2.045 (from Table C).
Note that the critical t uses N − 2 df

Since $1.66 < 2.045$, we retain H_0 and conclude that, based on our sample, there is not sufficient reason to conclude that the r of .3 that we obtained is different from zero. In other words, based on our data, we cannot be confident that there is any relationship between watching violent TV shows and aggression in play for 5-year-olds.

Notice that if we were to reject H_0, we reject the hypothesis that the population correlation is 0. This means that we can be confident that the population correlation coefficient for the variables in question is not zero. Rejecting H_0 does not allow us to say with a particular degree of confidence that the population correlation coefficient is the same as our sample correlation coefficient.

PREDICTION

As we described earlier, associated with every linear correlation coefficient is a best-fit straight line through the center of the scatterplot that is called a *regression line.* The criterion for this straight line is to minimize the sum of squared distances based on units of either the X axis or Y axis, as shown in Figure 10.8 using Y units. Because different regression lines will usually result from using X units as opposed to Y units, two regression lines can be associated with most scatterplots. However, we will simplify matters by showing only one regression line, based on minimizing Y distances. Should the need arise for the second regression line, a simple way to handle the situation will be presented. Because a regression line can be used for prediction, we will also call this line a *prediction line.* We will illustrate the basic idea involved in predicting by using an example of determining the relationship between practice and performance on a motor skills task. Suppose the task is "driving" in a simulated driving machine, such as is sometimes found in arcades. A score on this task can range from 0 to 5000, depending on the skill of the driver. We now sample 30 people who have played the game and obtain from them (a) the approximate number of previous games played (i.e., practice) and (b) their last score. Once we have this information, we can compute an *r* and a prediction line, as shown in Figure 10.9.

Figure 10.8
Linear regression based on *Y* units.

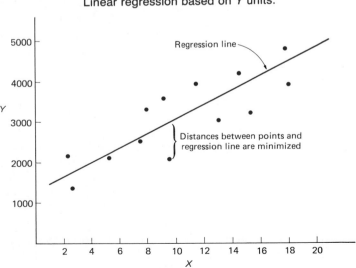

Figure 10.9
Scatterplot showing hypothetical relationship between driving performance and practice.

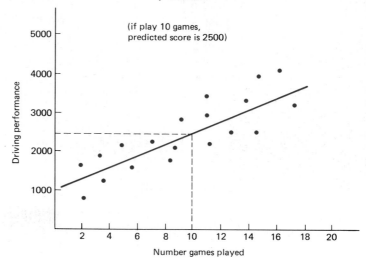

Based on the information in this scatterplot, two things are now possible: (a) knowing the value of one variable will allow us to make the best prediction possible concerning the value of the second variable. For example, if we see an unknown player's score, we can make a best possible prediction of how many games he or she has played. Also if we know how many games a person has played, we can make a best possible prediction of his or her next score. (In both cases, we assume that the individual record for the person in question is not available, since that would remove the variability between people, which is a part of the variability in the scatterplot in our example.) (b) Once we make a prediction, we can determine how accurate that prediction is likely to be. As you might suspect, the higher the *r*, the more accurate will be the prediction.

Let us look at two examples of how predictions can be made. Suppose you overhear a stranger say that this is the tenth time that he has played the game. Based on the prediction line in Figure 10.9, if we start at unit 10 on the *X* axis and draw a perpendicular line to the prediction line and then draw a horizontal line to the *Y* axis, we will have the predicted *Y* value. In this case, the value is 2500. How accurate our prediction will be depends on the distance from the prediction line to the points in the scatterplot at unit 10 on the *X* axis. For example, if scores for people having played 10 games range from 1500 to 3500, then, when we predict a score of 2500, we may make an error of 1000, but we are not likely to make an error of 2000.

Up to this point, we have discussed predictions and the magnitude of error associated with predictions in an informal way. However, techniques are available that will express both the regression (i.e., prediction) line and a measure of prediction error in terms of mathematical formulas.

PREDICTION LINES

The following represents the formula for a regression (prediction) line where we are predicting Y scores from X scores:

$$Y' = b_{yx}X + a_{yx} \qquad\qquad 10.4$$

where Y' = predicted score on Y
b_{yx} = slope of the line (also called the *regression coefficient* for predicting Y from X)
a_{yx} = Y intercept (the value of Y' when $X = 0$)

Note. **The subscript *yx* means predicting Y from X. A subscript *xy* means predicting X from Y.**

The formula for b_{yx} is

$$b_{yx} = r\frac{\sigma_y}{\sigma_x} \qquad\qquad 10.5$$

where r = correlation coefficient for a set of data
σ_y = standard deviation of the Y scores
σ_x = standard deviation of the X scores

In raw score format,

$$b_{yx} = \frac{N\Sigma XY - \Sigma X\Sigma Y}{N\Sigma X^2 - (\Sigma X)^2} \qquad\qquad 10.6$$

Equations 10.5 and 10.6 are equivalent. However, the raw score formula is easier to use under conditions where r, σ_x, and σ_y are not available from previous calculations or from a calculator programmed to do these functions.

The formula for a_{yx} is

$$a_{yx} = \overline{Y} - b_{yx}\overline{X} \qquad\qquad 10.7$$

where $\overline{Y}$ = mean of Y scores
b_{yx} = value in previous formula
$\overline{X}$ = mean of X scores

Let us now construct regression lines based on data from two of our original correlation examples, Situations b and c.

	Situation b		Situation c	
	X	Y	X	Y
S_1	19	14	19	14
S_2	18	15	18	9
S_3	17	13	17	11
S_4	15	11	15	15
S_5	14	10	14	10
S_6	13	9	13	13

$\bar{X}=16$ $\bar{Y}=12$ $\bar{X}=16$ $\bar{Y}=12$

$\sigma_x=2.16$ $\sigma_y=2.16$ $\sigma_x=2.16$ $\sigma_y=2.16$

$r=.96$ $r=-.11$

For Situation b,

$$b_{yx} = .96\left[\frac{2.16}{2.16}\right] = .96$$

$$a_{yx} = 12 - .96(16) = -3.36$$

Therefore,

$$Y' = .96X + (-3.36)$$

Since any two points define a straight line, we can choose any two values of X and find their corresponding Y values from the regression equation, plot them on a graph, and draw a straight line through them. This straight line is the regression line for predicting Y from X. For convenience, we choose $X = 0$ and $X = 10$. Thus

$$\text{when } X = 0 \quad Y' = .96(0) + (-3.36) = -3.36$$

$$\text{when } X = 10 \quad Y' = .96(10) + (-3.36) = 6.24$$

The scatterplot and regression line for predicting Y from X are shown in Figure 10.10, panel 1. Once the values for b_{yx} and a_{yx} have been determined, the regression formula itself can be used to predict Y from X without reference to a graph containing the regression line. Thus, in our example, if we wish to find Y' given an X value of 5, we would determine Y by using the regression equation as follows:

$$Y' = b_{yx}X + a_{yx}$$

$$Y' = .96(5) + (-3.36) = 1.44$$

Given an X value of 5, we would predict a Y value of 1.44.

Let us now compute a regression line for the data in Situation c. For these data, $\sigma_x = 2.16$, $\sigma_y = 2.16$, $\bar{X} = 16$, $\bar{Y} = 12$, and $r = -.11$.

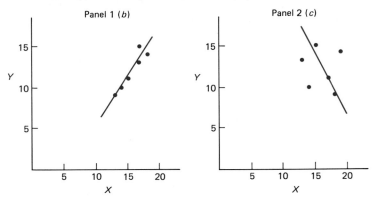

Figure 10.10
Scatterplot and regression line for Situations b and c.

Therefore,

$$b_{yx} = -.11 \left(\frac{2.16}{2.16} \right) = -.11$$

$$a_{yx} = 12 - (-.11)(16) = 13.76$$

The regression line for predicting Y from X for this set of data would be

$$Y' = -.11X + 13.76$$

This regression line and the corresponding scatterplot are shown in panel 2 of Figure 10.10. Notice how the points in this scatterplot are generally farther from the regression line than are the points in panel 1. This illustrates an important point: *the magnitude of prediction error decreases as the correlation coefficient gets larger.*

MEASURING PREDICTION ERROR: THE STANDARD ERROR OF ESTIMATE

standard error of estimate:

standard deviation of the distances from the points in a scatterplot to the regression line measured in units of x or y

We have seen how the further the points are from the regression line, the greater will be the prediction error. The most common way to represent this variability with a single measure is called the *standard error of estimate* (σ'_y). The standard error of estimate is computed just like the standard deviation for a set of scores, except that each score (with respect to its value on the Y axis) is subtracted from its predicted score (i.e., Y') represented by the regression line, rather than from the mean of the set of scores. Thus

$$\sigma'_y = \sqrt{\frac{\Sigma(Y - Y')^2}{N}}$$

10.8

An equivalent formula that is easier to work with is

$$\sigma'_y = \sigma_y \sqrt{1 - r^2}$$

10.9

Let us now compute the standard error of estimate σ'_y for the regression lines in Situations b and c. For Situation b, we found

$$\sigma_y = 2.16 \quad \text{and} \quad r = .96$$

Therefore,

$$\sigma'_y = 2.16 \sqrt{1 - (.96)^2} = .61$$

For Situation c,

$$\sigma_y = 2.16 \quad \text{and} \quad r = -.07$$

Therefore

$$\sigma'_y = 2.16 \sqrt{1 - (-.07)^2} = 2.15$$

As should be the case, we have a much larger value for σ'_y in Situation c than in Situation b. This means that predictions will be very good (i.e., small values for σ'_y) for Situation b, but very poor (i.e., large values for σ'_y) for Situation c.

REGRESSION OF X ON Y

If we want to compute a regression line based on the X axis instead of the Y axis, we simply interchange X and Y in all of the previous regression and standard error of estimate formulas. For example,

$$b_{xy} = r \frac{\sigma_x}{\sigma_y}$$

10.10

or

$$b_{xy} = \frac{N\Sigma XY - \Sigma X \Sigma Y}{N\Sigma Y^2 - (\Sigma Y)^2}$$

10.11

$$\sigma'_x = \sigma_x \sqrt{1 - r^2}$$

10.12

$$a_{xy} = \bar{X} - b_{xy}\bar{Y}$$

10.13

SUMMARY

The term *correlation* means relationship between variables. A *Pearson correlation coefficient* is a mathematical technique used to find a numerical value, ranging from -1 to $+1$, that measures the strength of the relationship between two variables. Independent variable and dependent variable labels are usually dropped when we talk about correlational variables. A large correlation coefficient does not imply that one variable is causing something to happen with respect to the other variable.

Data using the Pearson correlation coefficient technique are said to be *bivariate*; that is, each case has two variables associated with it. Thus, correlational data can be shown on graphs called *scatterplots*, wherein each case is a point on the graph representing a value of the X and Y variables. If the slope of the scatterplot is to the right, the scatterplot represents a positive correlation. If the slope is to the left, the scatterplot represents a negative correlation. If no slope is apparent, the correlation is zero or near zero. With high positive or negative correlations, there are only small deviations between the points in the scatterplot and the regression (i.e., best-fit) line through the points. With smaller correlations, distances between the points and the regression line increase.

The formula for finding a Pearson correlation coefficient is

$$r = \frac{\Sigma Z_x Z_y}{N}$$

A more convenient form that can be used with most calculators is the raw score formula

$$r = \frac{N\Sigma XY - (\Sigma X)(\Sigma Y)}{\sqrt{[N\Sigma X^2 - (\Sigma X)^2][N\Sigma Y^2 - (\Sigma Y)^2]}}$$

Properties of *r* are the following: (a) Adding or subtracting a constant (i.e., any single number) to or from the X or Y values does not change *r*. (b) Since both the X and Y scores are converted into Z scores before *r* is computed, the X and Y variables are converted into the same relative scale. (c) The correlation coefficient does not form a ratio scale, so we cannot say, for example, that an *r* of .4 shows a relationship that is twice as strong as an *r* of .2. However, r^2, which designates *percent of variance explained*, does constitute a ratio scale. Percent of variance explained reflects the extent to which knowing values of one variable will allow one to account for variability in the second variable. (d) The magnitude of *r* depends on the range of the X and Y variables. If the range of either variable is restricted, *r* will be lowered. If the middle of the range of either variable is omitted, *r* will be increased. (e) One can use *r* based on a sample to estimate the correlation coefficient of a population. The larger the number of cases in the sample correlation coefficient, the better the estimate will be of the population correlation coefficient.

The *r* from a sample can be tested to see if it is significantly different from zero by transforming *r* into a computed *t* score by the formula

$$t = \frac{r\sqrt{N - 2}}{\sqrt{1 - r^2}}$$

The critical t value is obtained from Table C in the appendix for $N - 2$ df.

A *regression line* allows one to make the best possible prediction of one variable based on values of the second variable. The formula for a regression line for predicting Y scores from X scores is

$$Y' = b_{yx}X + a_{yx}$$

where b_{yx} is the slope of the regression line, also called the *regression coefficient*, and a_{yx} is the Y intercept of the regression line. The formula for b_{yx} is

$$b_{yx} = r\frac{\sigma_y}{\sigma_x}$$

or

$$b_{yx} = \frac{N\Sigma XY - \Sigma X\Sigma Y}{N\Sigma X^2 - (\Sigma X)^2}$$

The formula for a_{yx} is

$$a_{yx} = \bar{Y} - b_{yx}\bar{X}$$

The most common way of computing the variability of points around the regression line is called the *standard error of estimate*, σ_y'. The standard error of estimate is computed just like the standard deviation for a set of scores except that, instead of subtracting each score from the mean of the set of scores, we subtract each score (with respect to its value on the Y axis) from its predicted score (Y') represented by the regression line. Thus,

$$\sigma_y' = \sqrt{\frac{\Sigma(Y - Y')^2}{N}}$$

An equivalent formula is

$$\sigma_y' = \sigma_y\sqrt{1 - r^2}$$

KEY TERMS

Pearson correlation coefficient

regression line

scatterplot

case

bivariate

percent of variance explained (r^2)

b_{yx}

a_{yx}

$Y' = b_{yx}X + a_{yx}$

σ_y'

positive slope

negative slope

restricting the range of r

b_{xy}

a_{xy}

PROBLEMS

1. For the following sets of data

    a   Plot the data as a scatterplot

    b   Compute r

    c   Compute b_{yx} (the regression coefficient)

    d   Compute a_{yx}

    e   Find Y' for any two values of X

    f   Draw the regression line on the scatterplot

    g   Compute σ'_y

	* Set 1			Set 2			Set 3	
	X	Y		X	Y		X	Y
S_1	13	6	S_1	35	20	S_1	0	12
S_2	15	8	S_2	21	34	S_3	8	1
S_3	10	1	S_3	33	27	S_3	10	5
S_4	15	6	S_4	25	30	S_4	7	9
S_5	9	2	S_5	23	24	S_5	12	7
S_6	11	7	S_6	29	27	S_6	1	2
S_7	14	7	S_7	28	28	S_7	9	0
			S_8	34	21	S_8	5	8
			S_9	31	27	S_9	4	11
			S_{10}	22	33	S_{10}	7	10
						S_{11}	6	6
						S_{12}	8	9

2. *a*   For set 2 in problem 1, add S_{11} values of $X = 21$ and $Y = 20$. Now compute r for the data.

   b   How does this r compare to the r that was calculated for set 2 in problem 1? What caused the change in the magnitude of r?

3. Using the data in problem 1, find the best prediction that one can make for Y (i.e., Y') using X values of 5 and 10 for

    *a   set 1

    b   set 2

    c   set 3

4. Test each of the r's in problem 1 for significance, using $\alpha = .05$ and a nondirectional test. Use the format in the example from the chapter.

5. Using the following X scores

	X	Y
S_1	10	
S_2	8	
S_3	6	
S_4	4	

 a Fill in any Y scores that would give an r of $+1$. Show that it is $+1$ by computing r.

 b Fill in any Y scores that would give an r of -1 and show that it is -1 by computing r.

 c Fill in any Y scores that would given an r of zero. [*Hint:* the numerator of $\Sigma Z_x Z_y$ must equal 0.] Compute r.

 d Draw scatterplots of the data in (a), (b), and (c).

6. a A researcher wished to test the hypothesis that girls of young mothers will themselves be young mothers. Using randomly selected women, she obtained data on the woman's age when she first gave birth and on the age of the woman's mother when the mother first gave birth. Find r and test it for significance using $\alpha = .05$ and a nondirectional test. Was the researcher's hypothesis supported?

	Woman's Age at First Birth	Woman's Mother's Age at First Birth
S_1	19	17
S_2	26	24
S_3	16	16
S_4	18	16
S_5	19	18
S_6	23	22
S_7	15	15
S_8	22	24
S_9	28	21
S_{10}	25	26
S_{11}	17	17
S_{12}	19	17
S_{13}	23	20
S_{14}	17	16
S_{15}	19	19
S_{16}	21	30
S_{17}	22	18

S_{18}	20	18
S_{19}	25	21
S_{20}	18	17
S_{21}	27	21
S_{22}	32	23
S_{23}	18	17
S_{24}	30	22
S_{25}	24	20

    b If a woman first gives birth at 24, what was the most likely age of her mother at first birth?

7. Test the following r's for significance, using $\alpha = .05$ and a nondirectional test.

    *a $r = .45, N = 10$
    b $r = .26, N = 70$
    c $r = .61, N = 5$
    d $r = .41, N = 30$

8. For each of the following situations, name two variables (not mentioned in the chapter) that you think would have a

    a high positive correlation
    b near zero correlation
    c high negative correlation
    d high correlation but no causal connection
    e high correlation with a causal connection. State why you think that the causal connection exists.

CHAPTER 11

Chi-Square

INTRODUCTION

Suppose we are interested in peoples' preferences of mental health professionals; in other words, given a free choice, do people prefer (a) psychiatrists, (b) psychologists, or (c) social workers? To obtain data on this question, we devise a scenario where each type of professional might provide appropriate services, and then ask randomly selected people whom they would choose. The question is now, ''How do we analyze the data that might be obtained from this experiment such that we might be confident that differences that we see in preferences represent real differences in a population and are not due to elements of chance that may exist within our sample? For example, if we poll 60 people, we may find that 25 would prefer a psychologist, 19 a psychiatrist, and 16 a social worker. Can we be confident, based on this data, that psychologists are really the preferred choice of the designated population from which our random sample was taken? Your first thought might be, "We have one independent variable with three conditions; let's use a one-factor ANOVA to analyze the data." The problem with this approach is that we have only one score, a frequency count, for each condition. With only one score per condition, we have no information on within-cell variability, and thus no way to compute an error term or to do a one-factor ANOVA.

categorical data:

data consisting of frequency counts within predetermined categories

However, as you might suspect, there is a way to perform our standard hypothesis testing procedure using data that are in the form of frequency counts within *categories.* The appropriate test for data in the form of frequency counts within categories is called *chi-square* (χ^2).

CHI-SQUARE TESTS FOR ONE VARIABLE

Let us now examine the logic of the χ^2 test. We start by asking what frequencies we would expect in a given set of conditions if the null hypothesis were true. We call these expected frequencies f_e. We then get a measure (called χ^2) of the difference between our observed frequencies (i.e., f_o) and expected frequencies (f_e) when H_0 is true by using the formula

$$\chi^2 = \Sigma \frac{(f_o - f_e)^2}{f_e}$$

11.1

where χ^2 = chi-square value
f_o = observed frequency count for a condition
f_e = expected frequency count for a condition
Σ = sum $(f_o - f_e)^2 / f_e$ for each condition

To compute a χ^2 value using our "type of therapist" example, we need to find f_e for each condition. Values for f_e are determined by noting that if the null hypothesis were true, then the expected frequency for each of the choices would be the same. Since there are three choices of therapist and 60 people making choices, the expected frequency for each choice is $(1/3)(60) = 20$. The expected and observed frequencies for our example are

	Psychiatrist	Psychologist	Social Worker
	$f_e = 20$	$f_e = 20$	$f_e = 20$
	$f_o = 19$	$f_o = 25$	$f_o = 16$

We can now compute a chi-square value as follows:

$$\chi^2 = \frac{(25 - 20)^2}{20} + \frac{(19 - 20)^2}{20} + \frac{(16 - 20)^2}{20}$$

$$= \frac{25 + 1 + 16}{20} = \frac{42}{20} = 2.1$$

The obvious question now is, what does a χ^2 of 2.1 mean? To answer this question, first notice that when f_o and f_e are similar in value, the computed χ^2 will be small. Conversely, when the difference between f_o and f_e is large, the value for χ^2 is large. The next question is, how large does χ^2 have to be before we can say that our observed frequencies are too discrepant from the expected frequencies to have occurred under conditions where the null hypothesis is true? We can answer this question using the same logic as we used with our other statistical tests of inference. We need to find the range of χ^2 values that are likely to occur when H_0 is true; if the χ^2 value for our example is beyond this range (i.e., exceeds the cutoff value), we reject H_0.

We could determine what χ^2 values to expect under conditions where H_0 is true in the following way. We could get one red, one blue, and one green marble and place them in a box. Each color of marble would represent a type of therapist. Now we shake the box, close our eyes, and draw a marble from the box and record the color. We then replace the marble, draw again, and record the color. We repeat this procedure for 60 draws. We then count the number of red, blue, and green marbles and use these counts as the f_o values in computing a χ^2 value. Suppose that we do several thousand repetitions of this whole procedure of computing χ^2 values from frequency counts of 60 randomly chosen red, blue, or green marbles. We will end up with a frequency distribution of χ^2 values that have been obtained under conditions where H_0 was really true (i.e., each marble was equally likely to be chosen on each draw). If we use $\alpha = .05$ as our decision criteria for rejecting H_0, we would find the χ^2 value that was exceeded for only 5% of the χ^2 numbers in our marble-drawing simulation. In other words, χ^2 values greater than this cutoff value would be rare when H_0 is really true. Therefore, if in our data concerning therapists, we obtain a χ^2 value greater than this cutoff χ^2 number, then we would reject H_0 (i.e., we would reject the notion that the frequency counts that we obtained occurred under conditions where each frequency was equally likely).

At this point, a very fortunate event happens. It turns out that the frequency distribution of χ^2 values for our marble-drawing experiment would look very similar to a theoretical distribution called the χ^2 distribution. Therefore, we do not need to find a critical χ^2 value from marble-drawing or other equivalent simulations; rather we can find a critical χ^2 value from a table based on the theoretical χ^2 distribution.

In summary, we use the χ^2 distribution as a model for finding cutoff values for making a decision about rejecting H_0 in the same manner as with the t and F distribution models that we discussed earlier. Before finding the critical χ^2 value appropriate for making a decision for our data concerning choosing therapists, let us examine in more detail the theoretical χ^2 distributions.

THE χ^2 DISTRIBUTIONS

The χ^2 distributions are a family of distributions that change their shape rather dramatically as a function of degrees of freedom. Degrees of freedom for χ^2 vary along one dimension (as with the t distributions) rather than two dimensions (as with the F distributions). However, the χ^2 distributions share more common characteristics with the F distributions than with the t distributions. This happens because both the χ^2 and the F distributions reflect squared values, which yield positive numbers only. Chi-square distributions for 1 and 10 df are shown in Figure 11.1.

Properties of the χ^2 distributions are the following:

(a)   They are a family of continuous distributions (i.e., smooth curves) that change shape with degrees of freedom.
(b)   They range in values from 0 to positive infinity.
(c)   They are *not* symmetrical and *not* normal, but become increasingly so on both measures with increasing degrees of freedom.

Figure 11.1
Chi-square distributions.

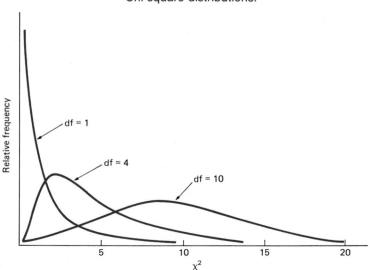

These characteristics of the χ^2 distribution are the same as for the F distribution. However, unlike the F distributions, the χ^2 distributions keep shifting to larger numbers along the x axis (see Figure 11.1) with increasing df, whereas the F distributions cluster around the number 1 with increasing df. To further illustrate this point, examine the tables containing critical (i.e., cutoff) F values and χ^2 values (Tables D and E). Notice that with larger df, the critical F values get smaller and the critical χ^2 values get larger.

DEGREES OF FREEDOM FOR ONE-VARIABLE SITUATIONS

For one-variable experiments, the degrees of freedom used to determine which theoretical χ^2 distribution is appropriate is found by taking the number of categories into which scores may fall and subtracting 1. Thus in our experiment, we have three categories representing types of therapist. Therefore, the appropriate χ^2 distribution would have $3 - 1 = 2$ df.

THE χ^2 TEST AND DIRECTIONALITY

The χ^2 test is usually considered to be nondirectional, although there can be exceptions. Let us use our example to illustrate. We are interested in seeing if large differences exist between the frequencies that we observe and what we would expect if the data were randomly placed into the designated categories representing types of therapists. We are not interested in whether these differences were too small to be random differences. Thus, if we use $\alpha = .05$, all of our rejection area would be placed in the χ^2 tail furthest along the x axis (i.e., the right and, in our example, the only tail). Because of the preponderance of nondirectional usage, χ^2 tables, such as Table E in the appendix, list critical values such that all of the rejection area is to the right of the critical χ^2 value. See Figure 11.2.

Figure 11.2

Illustration of the critical χ^2 value when df = 4.

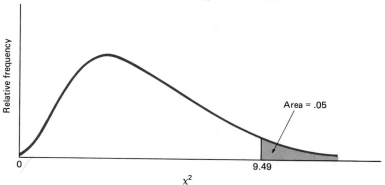

Relative frequency

Area = .05

0   9.49

χ^2

THE NULL HYPOTHESIS FOR χ^2 TESTS

The null hypothesis (H_0) for χ^2 tests can be stated in different ways. The following represents one basic way.

$$H_0: f_o \text{ differs from } f_e \text{ because of random influences}$$

In our example, if we reject H_0, we would infer that preferences for type of therapist are not the same within the population from which our sample was taken.

THE χ^2 TEST FOR OUR EXAMPLE

We are now in a position to answer our question concerning choice of therapist by using the customary hypothesis testing format. Using $\alpha = .05$, with 2 df, the critical χ^2 value from Table E in the appendix is 5.99. Thus, critical $\chi^2 = 5.99$, computed $\chi^2 = 2.10$. Since $2.10 < 5.99$, retain H_0. We therefore conclude that the frequencies of preferences that we obtained were too close to what would be expected if no preference existed for us to make any claim that one type of therapy is preferred over another.

USAGE OF χ^2 FOR ONE VARIABLE

A one-variable χ^2 test can be used in any situation where frequency counts can be obtained and expected frequencies determined. Thus one could empirically test whether or not a die was honest by rolling it a large number of times and doing a χ^2 test on the differences between the observed frequencies and the expected frequencies (obtained by dividing the number of rolls by 6).

Since frequency distributions involve frequency counts, the shape of an empirically obtained frequency distribution can be tested using χ^2 to see if it matches the shape of a predetermined distribution. For example, suppose there exists a predetermined frequency distribution such that 60% of people favoring increasing foreign aid will be male and 40% female. Suppose we polled 100 people who favored foreign aid. We would expect 60 males and 40 females in our sample. If the actual data were 50 and 50, we could compute χ^2 as follows:

$$\chi^2 = \frac{(50 - 60)^2}{60} + \frac{(50 - 40)^2}{40} = 4.17$$

$$\text{critical } \chi^2, \quad (\alpha = .05, \quad 1\,df) = 3.84$$

Since $4.17 > 3.84$, the predicted distribution of frequencies is rejected.

TWO-VARIABLE SITUATIONS: χ^2 TESTS FOR INDEPENDENCE

AN EXAMPLE

independent:

not influencing or being influenced by anything else

The most common usage of χ^2 tests is to see if categories representing two variables are related. If they are not related, we say that the two variables are *independent* of one another. Let us illustrate the χ^2 procedure for the two-variable situation within the context of an example. Suppose we wish to study the effects of an up-mood state (i.e., "feeling good") on helping. To induce this "feeling good" state, we place two quarters in the change slot of a candy machine where normally a person would expect to get 5 or 10 cents change after a purchase. Shortly following the subject's "finding" the extra change, the experimenter "accidentally" drops several papers in the vicinity of the subject. The dependent variable is whether the subject offers to help pick up the papers. The same situation also occurs for control subjects who do not find extra change in the machine. Thus one variable (the independent variable) involves finding or not finding change, and the other variable (the dependent variable) involves helping or not helping. When the variables are shown combined into a 2 × 2 matrix, each subject will have a mark (i.e., tally) in one and only one of the four resulting cells. Suppose the data were distributed as shown in Table 11.1:

TABLE 11.1
Data for the Hypothetical Study of
Effects of Feeling Good on Helping

		Helping			
		Yes	No		
Finding Money	Yes	40	10	50	
	No	20	30	50	
	Total	60	40	100	Grand Total

The row 1 total of 50 means that 50 people found extra money in the machine. Within this number 40 helped and 10 did not. Row 2 indicates that of the 50 people who did not find extra money, 20 helped and 30 did not. A similar type of breakdown can be made for columns 1 and 2. Of the 60 people who helped, 40 found money. Of the 40 people who didn't help, 10 found money.

Now if helping is not related (i.e., independent of) to finding money, what would we expect as frequencies within the cells? The answer is that we would expect the same proportion (i.e., percentage) of helping for both the find-money group and the do-not-find-money group. A more general way of saying this is that if two variables are not related, they will have the same (within random variations) proportions within each row and within each column. Thus the expected frequencies will have this property.

To find the expected frequency for each cell, take the row total for the row that contains that cell times the column total for that cell and divide by the overall total. Symbolically, this is written

$$f_e = \frac{\text{row total} \times \text{column total}}{\text{overall total}} \qquad 11.2$$

Computing the expected frequencies, we have

$$f_e \text{ for cell 1 (row 1, column 1)} = \frac{50 \times 60}{100} = 30$$

$$f_e \text{ for cell 2 (row 1, column 2)} = \frac{50 \times 40}{100} = 20$$

$$f_e \text{ for cell 3 (row 2, column 1)} = \frac{50 \times 60}{100} = 30$$

$$f_e \text{ for cell 4 (row 2, column 2)} = \frac{50 \times 40}{100} = 20$$

Placing these values for expected frequencies within our 2×2 matrix, we can see that the same proportions are maintained within rows and columns:

		Helping	
		Yes	No
Finding Money	Yes	30	20
	No	30	20

The question that we now ask is, "Are the frequencies that we have observed in our experiment different enough from what would be expected if the variables were not related for us to reject H_0 (i.e., the not-related hypothesis)?"

As you might suspect, the null hypothesis for tests of independence is

H_0: the two variables (X and Y) are not related or, equivalently, the variables are independent

Thus, H_0 for our experiment would be

H_0: finding money and helping are not related

To test whether the frequencies that we have observed in our experiment deviate far enough from the expected frequencies if H_0 were true, we compute a χ^2 value as follows:

$$\chi^2 = \Sigma \frac{(f_o - f_e)^2}{f_e}$$

	f_o	f_e	$f_o - f_e$	$(f_o - f_e)^2$	$(f_o - f_e)^2/f_e$
Cell 1	40	30	10	100	3.33
Cell 2	10	20	10	100	5.0
Cell 3	20	30	10	100	3.33
Cell 4	30	20	10	100	5.0
					$\Sigma = 16.67$

Our computed $\chi^2 = 16.67$.

DEGREES OF FREEDOM FOR χ^2 TESTS OF INDEPENDENCE

Before we can determine if our computed chi-square value of 16.67 is large enough to reject H_0, we need to find the appropriate df for our χ^2 theoretical distribution. This is found by taking (as we did for interaction in two-factor ANOVA) the number of rows (R) minus 1 times the number of columns (C) minus 1:

$$df(\text{independence tests}) = (R - 1)(C - 1)$$

Thus, in our example,

$$df = (2 - 1)(2 - 1) = 1$$

Notice that in χ^2 tests, unlike the t and F tests that we have studied, the df for the critical χ^2 value is *not* influenced by the total number of subjects nor scores (i.e., tallies), but is determined by the number of categories.

SUMMARY OF THE χ^2 TEST FOR INDEPENDENCE

We are now in a position to draw some conclusions from our example study. To summarize:

H_0: the two variables (helping and "finding" money) are not related.

Critical χ^2 (1 df) (using $\alpha = .05$) = 3.84

Computed $\chi^2 = 16.67$

Since $16.67 > 3.84$, reject H_0.

We conclude that finding money is related to helping. A typical way of reporting these χ^2 results in a journal would be as follows: "A 2×2 chi-square test for independence showed a significant relationship between helping and finding money, χ^2 (1 df, $N = 100$) = 16.67, $p < .05$." Notice that this test only allows us to say something about the independence of the specific variables that we used in the study. The statistical test does not by itself demonstrate that finding money made people "feel good" and that "feeling good" was the reason for increased helpfulness (see again Chapter 1 and Table 1.2). The interpretation that finding money makes a person feel good is a separate inference that must be made.

EXAMPLE 2: χ^2 TEST FOR INDEPENDENCE

An investigator is interested in the effect of different types of appeals on volunteering to give blood to the Red Cross. Three appeals are devised: (a) negative emphasis: what will happen if not enough blood is donated?; (b) positive emphasis: the benefits of giving blood; (c) combination of negative and positive emphases. Each appeal is given to a large section of an introductory psychology course, and students are then asked to volunteer. The hypothetical data are given in Table 11.2:

TABLE 11.2
Data for Example 2

		Positive	Negative	Combined	
	Yes	30	20	50	100
Volunteer	No	60	90	50	200
	Total	90	110	100	300

Type of Appeal

Is type of appeal related to volunteering to give blood? An identical way of asking this question is, "Is the proportion of volunteers the same for all types of appeals?"

H_0: type of appeal is not related to volunteering to give blood.

df $= (R-1)(C-1) = (1)(2) = 2$

$\alpha = .05$,

Critical $\chi^2 = 5.99$ (from Table E)

To calculate χ^2, we find f_e for each cell, we have

$$\text{cell 1 (row 1, column 1): } \frac{90 \times 100}{300} = 30$$

$$\text{cell 2 (row 1, column 2): } \frac{110 \times 100}{300} = 36.67$$

$$\text{cell 3 (row 1, column 3): } \frac{100 \times 100}{300} = 33.33$$

$$\text{cell 4 (row 2, column 1): } \frac{90 \times 200}{300} = 60$$

$$\text{cell 5 (row 2, column 2): } \frac{110 \times 200}{300} = 73.33$$

$$\text{cell 6 (row 2, column 3): } \frac{100 \times 200}{300} = 66.67$$

We now find

$$\sum \frac{(f_o - f_e)^2}{f_e}$$

	f_o	f_e	$f_o - f_e$	$(f_o - f_e)^2$	$(f_o - f_e)^2/f_e$
Cell$_1$	30	30	0	0	0
Cell$_2$	20	36.67	-16.67	277.89	7.59
Cell$_3$	50	33.33	$+16.67$	277.89	8.34
Cell$_4$	60	60	0	0	0
Cell$_5$	90	73.33	$+16.67$	277.89	3.79
Cell$_6$	50	66.67	-16.67	277.89	4.19

$$\sum \frac{(f_o - f_e)^2}{f_e} = 23.91$$

$$\text{Computed } \chi^2 = \sum \frac{(f_o - f_e)^2}{f_e} = 23.91$$

Since $23.91 > 5.99$, reject H_0. We conclude that type of appeal is related to volunteering to give blood. From the data, we see that the most effective appeal was the combined appeal, and the least effective appeal was the negative appeal.

FOLLOW-UP TESTS FOR χ^2

As was the case with ANOVA, when a design having several conditions is significant, an investigator may want to know which conditions within the design are significantly different from one another.

Since the data within cells in a chi-square design are independent, one can break the design into smaller segments and test the resulting comparisons. It is possible to do as many independent tests as there are df in the design.

To illustrate, in our blood volunteer example we may want to know if positive appeals differ from negative appeals, or perhaps if positive appeals differ from the combination of negative appeals with the dual (positive and negative) condition. If an investigator decided to test the positive appeal against the negative appeal in our example, he or she would perform a 2×2 test using the appropriate cells. In our example this would be

Appeal

		Positive	Negative
Volunteer	Yes	30	20
	No	60	90

The investigator's interests and judgment determine which follow-up tests are appropriate. A conservative procedure to use for follow-up tests is to perform them only following a significant χ^2 test. The reason for recommending a conservative approach to the follow-up tests is (as we discussed in Chapter 7) that performing more tests increases the probability that somewhere H_0 will be incorrectly rejected.

SHORTCUT PROCEDURE FOR 2 × 2 CHI-SQUARE MATRICES

For χ^2 tests of independence having two rows and two columns, there is a shortcut procedure. We label the numbers (i.e., frequency counts) within the four cells in a 2×2 matrix as follows:

		Row Totals
A	B	$A + B$
C	D	$C + D$

Column Totals $A + C$ $B + D$ $N =$ Grand Total

Then

$$\chi^2 = \frac{N(AD - CB)^2}{(A+B)(C+D)(A+C)(B+D)} \qquad 11.3$$

To illustrate, we will compute the data from our helping experiment shown in Table 11.2:

		Row Totals
40	10	50
20	30	50

Column Totals   60   40 $N = 100$

Thus,

$$\chi^2 = \frac{100[40 \times 30) - (20 \times 10)]^2}{50 \times 50 \times 60 \times 40} = \frac{100(1000)^2}{50 \times 50 \times 60 \times 40} = 16.67$$

This value is the same as we obtained with the longer method.

SOME PRECAUTIONS WHEN USING THE χ^2 TEST

The χ^2 test assumes that every frequency tally is independent of every other frequency tally. This means that a subject should contribute one and only one tally within the χ^2 matrix. For example, it would not be proper to compare attitudes on foreign aid between the sexes by giving a 10-item questionnaire concerning foreign aid to each of 10 males and 10 females and then tallying each response from the questionnaire within the χ^2 matrix to give a total N of 200. Rather, the questionnaire itself should be evaluated to yield one tally, yielding a total N of 20.

The appropriateness of the χ^2 theoretical distribution as a model from which to analyze data in the form of frequency counts depends on the size of the expected frequencies within the cells of the χ^2 matrix. The χ^2 test using 1 df is the most seriously distorted when small expected frequencies (i.e., f_e) are present. A rule of thumb is not to use the χ^2 test for 1 df if the expected frequency of any of the cells is less than 5. For χ^2 tests using 2 df, all expected frequencies should be greater than 2. With 3 df all expected frequencies should be greater than 1. With 4 or more df these expected frequency constraints do not apply. Since expected frequencies will get larger as the overall N gets larger, problems involving small expected frequencies can be overcome in situations where large sample sizes can be obtained.

CHI-SQUARE TESTS AND POWER

Unlike the t and F tests that we have studied, a chi-square test is not based on assumptions about population parameters, and is therefore called a *nonparametric* test (see Chapter 12 for a discussion of the assumptions underlying the t and F tests). As we saw with the sign test, and is generally the case, nonparametric tests are not as powerful as their parametric counterparts (e.g., the t test). By this we mean that when H_0 is false, a parametric test will reject H_0 more often than will a nonparametric test. Although there are no parametric equivalents to chi-square tests for independence, the idea of lack of power shows up in situations where a change in the dependent variable would allow a parametric test to be performed.

To illustrate, in many cases the dependent variable is inherently categorical, and the investigator has no choice but to use χ^2. For example, a person will or will not volunteer to give blood or to vote a particular way on a proposal. However, in other situations the experimenter may have a choice. To illustrate, suppose that the dependent variable is helping. This might be defined as the amount that one is willing to help, not just whether or not one helps. To demonstrate within a broader context, suppose that one is interested in the effects of different types of appeals on helping. If helping were defined as the number of hours volunteered to help by a subject instead of whether or not a subject helped, then the data could be analyzed within the context of a one-factor ANOVA instead of using a χ^2 test. The use of an ANOVA would allow a considerable savings in the number of subjects necessary for the experiment while keeping the *power* of the test the same. (Equivalently, given the same number of subjects, the ANOVA test is more powerful.) The magnitude of the subject savings can't be precisely determined without knowing the details of the situation involving the variability of the dependent variable. However, a ballpark figure would be the need for two to three times as many subjects for the χ^2 test compared with the ANOVA test.

STRENGTH OF ASSOCIATION WITH TWO-VARIABLE χ^2

The null hypothesis for the χ^2 tests of independence states that the two variables being tested have zero relationship with each other. If H_0 is rejected, we can say that some relationship exists between the variables, but the magnitude of this relationship is not specified. An index of the magnitude of the relationship between the two variables, such as the Pearson r that we discussed in the last chapter, would therefore be useful.

In fact, with χ^2 data in the form of a 2 × 2 matrix, we could compute r if we assign numbers to the categories representing the two variables. For example, let's look again at our 2 × 2 chi-square design involving ''finding'' money and helping. If we arbitrarily assign the numbers 0 and 1 to the ''not-finding'' and ''finding'' categories and again to the ''did not help'' and ''helped'' categories, we have bivariate data (i.e., each person has two scores) and thus a Pearson r can be computed. Since the manner of assigning numbers to categories is arbitrary, the sign of r would be arbitrary, but the magnitude of r would still show the strength of association. The computations for computing r would follow exactly those presented in the last chapter.

However, there is a shortcut for finding r in this 2×2 chi-square situation, provided that a value for χ^2 has been computed. This short-cut correlation index is called the *phi coefficient* (ϕ) and is computed as follows:

$$\phi = \sqrt{\frac{\chi^2}{N}}$$

11.4

where χ^2 is the value computed from the data and N is the total number of tallies.

The value for ϕ will range from 0 to 1, and is interpreted as a Pearson r without a sign. The direction of the relationship is determined from the matrix containing the data. Note that ϕ is usually not computed unless the χ^2 test is significant, since retaining H_0 implies that no meaningful relationship exists between the variables.

Example In our hypothetical study involving "finding" money and helping, we obtained a χ^2 value of 33.33, which allowed us to reject H_0 at the .05 level. The N in this example was 100. Therefore,

$$\phi = \sqrt{\frac{33.33}{100}} = .57$$

A correlation of this size indicates that a strong relationship exists between finding money and helping.

Cramer's ϕ

The phi coefficient can only be used for data within 2×2 matrices. However, a more general version of ϕ exists that can be used for any size matrix. This index is called *Cramer's ϕ* and is computed as follows:

$$\text{Cramer's } \phi = \sqrt{\frac{\chi^2}{N(k-1)}}$$

11.5

where χ^2 = value obtained from data
  N = total number of tallies
  k = the smaller number of rows or columns
    in the χ^2 matrix

Cramer's ϕ is interpreted exactly as the 2×2 phi coefficient and is the same in the 2×2 situation.

We will illustrate Cramer's ϕ, using the data from our example experiment on the relationship between type of appeal and volunteering to give blood. This was a 2×3 chi-square design that yielded a significant χ^2 value of 23.91 and contained a total N of 300. Therefore,

$$\text{Cramer's } \phi = \sqrt{\frac{23.91}{300(2-1)}} = \sqrt{\frac{23.91}{300}} = .28$$

THE CONTINGENCY COEFFICIENT C

The student should be aware of another measure of strength of association available for use following χ^2 tests. This is the contingency coefficient C, computed as follows:

$$C = \sqrt{\frac{\chi^2}{N + \chi^2}}$$

11.6

The value of C can vary from 0 to 1, but it is not a correlation coefficient and is therefore harder to interpret. It has other undesirable characteristics that cause statisticians to prefer the use of Cramer's ϕ instead of C. However, C has often been reported in journal articles, so you should be aware of its existence.

SUMMARY

The chi-square test is appropriate for data in the form of frequency counts within categories. The formula for computing the χ^2 statistic is

$$\chi^2 = \Sigma \frac{(f_o - f_e)^2}{f_e}$$

where f_o is the observed frequencies within categories and f_e is the expected frequencies within categories under conditions where the null hypothesis is true.

The χ^2 distributions are a family of distributions that have the following properties: (a) they change their shape as a function of degrees of freedom along one dimension; (b) they range in values from 0 to positive infinity; (c) they are *not* symmetrical and *not* normal, but become increasingly so on both measures with increasing degrees of freedom.

The χ^2 test is usually considered to be nondirectional, meaning that the experimenter is interested in determining whether discrepancies between observed and expected frequencies are too large to have occurred by chance. Thus, all of the H_0 rejection area occurs in the right tail of the χ^2 distribution. The most basic way of stating the null hypothesis for χ^2 tests is that H_0: f_o differs from f_e because of random influences.

Chi-square tests involving one-variable situations require determining expected frequencies on some predetermined basis such as that all expected frequencies be the same. Chi-square tests for two-variable situations are called tests for independence. Expected frequencies for χ^2 tests of independence have the property of equal proportionality across row and column conditions. Degrees of freedom for the one-variable case are determined by the number of conditions minus 1, for the two-variable case they are determined by the number of rows minus 1 times the number of columns minus 1. In χ^2 tests, the df for the critical χ^2 is determined by the number of categories, not by the number of the frequency count within the categories.

Assumptions of the χ^2 test that necessitate precautions when using the test are two: (a) each frequency tally should be independent of every other frequency tally; and (b) certain minimum expected frequencies need to be maintained for each cell, particularly when less than 4 df are involved.

Chi-square tests are not as powerful as t tests or ANOVA tests. Often, data are inherently categorical by nature and χ^2 tests must be used. Sometimes, however, a dependent variable that appears to be categorical can be placed on a continuous scale having several points, and t or F tests can be used to analyze the data.

When significance is obtained in two-variable χ^2 tests, a strength of association measure can be obtained. The preferred measure is called *Cramer's ϕ* and is computed as follows:

$$\text{Cramer's } \phi = \sqrt{\frac{\chi^2}{N(k-1)}}$$

where χ^2 = value obtained from data
  N = total number of tallies
  k = the smaller number of rows or columns
    in χ^2 matrix

Another measure of association, the *contingency coefficient,* is not recommended.

KEY TERMS

categorical data	one versus two-variable χ^2
f_o	strength of association
f_e	the phi coefficient
N	Cramer's ϕ
independence	contingency coefficient

PROBLEMS

(Use $\alpha = .05$ for all problems.)

*1. When a die was rolled 120 times, the following frequency count of the six numbers came up. Was the die honest?

Number on face of die

	1	2	3	4	5	6
Frequency	29	21	15	20	19	16

2. A coin was flipped 200 times. Heads came up 110 times and tails 90 times. Was the coin honest?

3. A manufacturer claims that his product X is preferred by twice as many people as product Z. A random sample of 900 people showed that 550 people preferred product X and 350 people preferred product Z. Was the claim of the manufacturer of product X justified?

4. The human relations department of a university is interested in knowing whether the proportions of males and females are the same across classifications of professors. Can equivalent proportions be assumed?

	Assistant Professor	Associate Professor	Full Professor
Males	63	71	43
Females	38	40	19

5.* A psychologist wants to know if the appearance of a defendant will influence a jury's decision in a mock trial situation. Pictures of "defendants" that were rated high, medium, or low on an appearance scale were taped on a sheet of paper describing an alleged burglary. The story of the crime was the same for each type of picture of a defendant. Different judges in each condition were used.

   a   Was conviction related to appearance?

		Low	Medium	High
Decision	Guilty	22	15	11
	Innocent	8	15	17

Appearance (header over Low, Medium, High)

   b   Compute the strength of the association.

6. An investigator designated one nonsense syllable in the middle of a list of 15 nonsense syllables to be used to compare the efficiency of mediators versus no mediators on recall. Subjects were given the list of nonsense syllables one at a time at a constant rate and told to tell the investigator if any of the nonsense syllables reminded them of a word or meaningful concept. The investigator recorded a yes or no for each nonsense syllable, but analyzed only the target syllable for each subject to preserve the assumption of independence of the tallies. The dependent variable was whether the target syllable was recalled when

the subject was instructed to try to recall all nonsense syllables following presentation of the list. The data for 60 subjects were as follows:

Recall

	Yes	No
Association to Syllable	22	10
No Association to Syllable	6	22

   a Use both the long and shortcut methods to determine whether a significant relationship exists between associating meaning to nonsense syllables and the recall of the syllables.

   b Compute a phi coefficient and comment on its size.

7. An investigator wants to know if political preference varies as a function of year in college. Based on the following preference data from students, is year in college related to political preference?

Year in College

Political Preference		1	2	3	4
	Democrat	22	18	19	15
	Republican	20	20	15	13
	Independent	8	12	16	22

8. Based on brain laterality effects, an investigator predicts that right-handers will do better at solving a certain type of mechanical puzzle than left-handers. Was her hypothesis supported? Also, if necessary, do a degree of association test.

Solved Problems

	Yes	No
Right-handers	48	52
Left-handers	27	73

9. Compute a chi-square value for the following data:

a

10	6
6	10

b

50	30
30	50

c Is the proportional difference between the four cells in (a) the same as in (b)?

d What does this problem tell us about the effect of sample size (i.e., N) on computed chi-square values?

10. Given the following situations, fill in any numbers that will sum to both the row and column totals that are given.

a

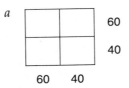

60
40

60   40

b

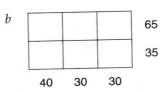

65
35

40   30   30

c In situation (a), for how many cells were you free to choose any number to place within that cell? In situation (b), for how many cells were you free to choose any number to place within that cell?

d How does the number of cells within which you could place any number compare to df as computed by $(R-1)(C-1)$? [This is why df is computed as $(R-1)(C-1)$.]

THOUGHT QUESTIONS

1. In what ways is the two-variable χ^2 test similar to the interaction test for a two-factor ANOVA?

2. Suppose that the H_0 rejection region for problem 4 was divided such that half was in each end of the χ^2 distribution. How might the interpretation for a computed χ^2 value that falls in the left rejection area (i.e., near zero) be different than one that falls in the right rejection area?

3. Change the dependent variable in problem 5 in such a way that one could do an ANOVA on the data. Would the appropriate ANOVA involve one factor or two factors? Explain.

CHAPTER 12

Special Statistical Topics

ASSUMPTIONS OF *T* AND *F* TESTS

As we pointed out in Chapter 5, theoretical distributions such as the normal distribution, the t distribution, and the F distribution are mathematically defined and have specific shapes. The t and F distributions are sometimes called theoretical sampling distributions, because using the t distribution as an example, when equal sized samples are taken from a normally distributed population, properties of these samples (e.g., differences between pairs of means) will form a frequency distribution having the shape of a t distribution.

In the theoretical model, the samples are taken from a normal distribution (or from two normal distributions having equal Variances). Therefore, we must assume that real-world samples also come from a normally distributed population in order for the t distribution to be an appropriate model with which to test a null hypothesis. If the normal distribution assumption is seriously violated, the resulting sampling distribution will not follow the shape of a t distribution.

Thus, to use that distribution as a model from which to make a decision about differences between sample means will result in an error. The magnitude of the error will depend on the degree to which the normal distribution assumption is violated.

Since empirical populations never *exactly* match a normal distribution (although they may be very close), the question becomes one of the magnitude of the violation. If the violation is small, the t distribution is a good model from which to make inferences about characteristics of populations. One way to guarantee that the violation of the normal distribution assumption will be small is to use large sample sizes. The reason is that sample means (the usual characteristic of samples that is of interest) will become increasingly normally distributed as N (i.e., the number of scores in a sample) increases, regardless of the shape of the population from which they were taken. This principle is called the *central limit theorem* (see Chapter 5). Therefore, if each condition in an experiment has a sample size of about 20 or more, any violation of the normal distribution assumption of the t test will be small enough to allow this violation to be effectively ignored.

With empirical data, we must also assume that the population variances estimated by our two sample variances are equal, because the theoretical model assumes that samples are taken from the same normally distributed population (or from two normally distributed populations having equal Variances). This violation can also be minimized with sample sizes of 20 or more. Thus, we can say that the t test is **robust** and is a good model even when the theoretical assumptions are not met, when large sample sizes are used.

The same assumptions that hold for the t test are also true for the F test for the same reasons. As with the t test, the F test is also robust with large sample sizes of 20 or more per condition. The F test is also more robust when the N's per condition are identical. Planning an experiment with equal N's of 20 or more per condition will minimize concerns about assumptions underlying the t or F tests.

robust :

(as in robust test) a statistical test that will give accurate results despite violation of the assumptions within the theoretical model underlying the test

SAMPLE SIZE ISSUES

STATISTICAL SIGNIFICANCE VERSUS PRACTICAL SIGNIFICANCE

Statistical tests of inference such as t tests or F tests are designed to allow the researcher to retain or reject the null hypothesis (H_0). The null hypothesis always takes the form of an equality (for example, $\mu_1 = \mu_2$). When one rejects the null hypothesis, one can be confident (at $1 - \alpha$) that the equality is not true (for example, that $\mu_1 = \mu_2$ is not true). However, the magnitude of the inequality is not specified. Thus, 100 is not equal to 800, and is also not equal to 100.1. Either of these comparisons conceivably could be statistically significant, although the magnitude of the differences between these pairs of numbers is vastly different. *Thus, finding a comparison to be statistically significant (i.e., H_0 is rejected) does not necessarily mean that this comparison has any practical significance or relevance.* One should not be misled by this statement. Most studies showing statistical significance probably also have some practical significance or relevance. The point is, however, that this does not always have to be the case. A comparison can be statistically significant and have no practical significance or importance.

THE EFFECT OF SAMPLE SIZE ON REJECTING H_0

In any and all statistical tests of inference, the probability of rejecting a false H_0 increases as sample size (N) increases. Let us see how this principle works by using the t test as an example. From Chapter 6, the formula for the t test having equal N in each condition is

$$t = \frac{\overline{X}_1 - \overline{X}_2}{\sqrt{S_1^2/N_1 + S_2^2/N_2}}$$

Now, assume that we are comparing fifth-grade boys to fifth-grade girls on a motor skills task where the scores can range from 0 to 200. We will now compute t under three different circumstances where everything in the t formula is the same except for N_1 and N_2. In our examples, boys have a mean score of 100.3, and girls have a mean score of 100.1. In our examples, $S_1^2 = 100$ and $S_2^2 = 100$.

Case 1 $N_1 = N_2 = 100$

$$t = \frac{100.3 - 100.1}{\sqrt{100/100 + 100/100}} = \frac{.2}{\sqrt{2}} = .14$$

Case 2 $N_1 = N_2 = 1000$

$$t = \frac{100.3 - 100.1}{\sqrt{100/1000 + 100/1000}} = \frac{.2}{\sqrt{.2}} = .44$$

Case 3 $N_1 = N_2 = 100,000$

$$t = \frac{100.3 - 100.1}{\sqrt{100/100,000 + 100/100,000}} = \frac{.2}{\sqrt{.002}} = 4.44$$

Testing each of these cases for significance at $\alpha = .05$, we see that Cases 1 and 2 are not significant, but Case 3 is significant. Since the standard deviation (e.g., S) for both boys and girls is 10, a mean difference of .2 is totally trivial, yet in Case 3 this difference was statistically significant. Although N's of 100,000 are unusual and often difficult to obtain, in this situation, an N of this size is conceivable if one had access to school records of students throughout a state or large city school system.

Interpreting statistical tests that are based on different sample sizes is a very important problem for the researcher. Closely associated to this problem is the choice of a sample size for a particular experiment. We will now discuss three different types of solutions to problems involving sample size: (a) power analysis concepts; (b) percent of variance accounted for; and (c) point estimates and confidence intervals.

SOLUTIONS

Power Analysis Concepts

Power analysis is a technique that can be used to equate statistical significance with practical significance such that any results that are statistically significant (i.e., H_0 is rejected) will also be of practical significance (i.e., differences between comparisons will be large enough to be meaningful). *Power is defined as the probability of correctly rejecting H_0 in a statistical test* and is equal to $1 - \beta$, where β is the probability of a Type II error (see Chapter 4). Power is a function of the significance criterion α, the effect size (e.g., difference between means), and the sample size N. Formal techniques of power analysis will not be presented here although they are available in other books (e.g., Welkowitz, Ewen, and Cohen, 1982). Instead, we will informally present enough of the concepts associated with power analysis for you to understand why techniques of this type are useful.

Again, let us use the t test as a model to illustrate the principles associated with power analysis. The size of a t statistic (i.e., computed t value) depends on three variables: (a) the difference between means, (b) the Variance (i.e., S^2) within the conditions, and (c) the sample size (i.e., N) of each condition. One purpose of formal power analysis as well as the technique herein described is to choose the size that N_1 and N_2 must be in order that H_0 will be rejected only when the difference between $\bar{X}_1$ and $\bar{X}_2$ is large enough to be meaningful. However, before the appropriate N's can be chosen, the experimenter must (a) decide how large the difference between $\bar{X}_1$ and $\bar{X}_2$ must be to be meaningful in a practical sense, and (b) make an estimate of the Variance (i.e., S_1^2 and S_2^2) that will be involved in the situation. Once values (a) and (b) have been estimated, we can solve the t formula for N to find out how many subjects should participate in the experiment such that a meaningful difference between means will also be statistically significant. Since one should plan for N_1 to equal N_2, we will let N represent both N_1 and N_2. The size of a critical t (i.e., cutoff value for retaining or rejecting H_0) varies with N (actually degrees of freedom), which is unknown but is usually in the 2.0–2.5 range for the standard nondirectional test using $\alpha = .05$. Therefore, we will use a middle range t value of 2.3 in the equation. The t formula is

$$t = \frac{\bar{X}_1 - \bar{X}_2}{\sqrt{S_1^2/N_1 + S_2^2/N_2}}$$

Solving for N, we have

$$N = \frac{t^2(S_1^2 + S_2^2)}{(\overline{X}_1 - \overline{X}_2)^2}$$

12.1

To illustrate how this works, let us use our previous example comparing fifth-grade boys and girls on a motor skills task. Suppose we decide that a meaningful difference between boys and girls is 5. We then estimate that S_1^2 (boys) and S_2^2 (girls) will be the same with a value of 100. The sample size that we would need to use to find statistical significance *if* there is a meaningful difference between the mean scores for boys and girls is

$$N = \frac{2.3^2(100 + 100)}{5^2} = 21.16$$

Therefore, if we use 21 subjects per condition and find statistically significant results, the difference between means will also be meaningful. Further, if we fail to find statistical significance, it is likely that any differences that may exist between the two groups are not meaningful.

Evaluation   Although techniques of the type described here offer a viable solution to the problem of incongruity between statistical and meaningful significance, they have their own set of problems. First, with data where several statistical tests are made on different combinations of the data, such as with a two-factor ANOVA, different comparisons may require different N's per cell to equate statistical significance with practical significance. The problem is further compounded with mixed designs wherein two different error terms are used. The comparisons using the within-subject error term are usually much more powerful (i.e., higher probability of correctly rejecting H_0) for any given sample size. Thus, the required N's for equating statistical and practical significance for the between and within comparisons are likely to be quite different.

A second problem with techniques such as the one described is that it requires a commitment on the part of the investigator as to what constitutes a meaningful outcome. Often, an investigator would like to use a broader scale for the meaningfulness dimension than one that has only two points, a yes and a no. For example, one might think that a 3-point difference between means is somewhat meaningful, a 5-point difference is rather meaningful, a 7-point difference is meaningful, and a 10-point difference is very meaningful. This type of gradation of the meaningfulness scale is not possible with the described technique.

A third problem with techniques such as the one described is consistency across investigators. A meaningful outcome for one investigator may not be a meaningful outcome for another. Despite the aforementioned problems, an investigator will usually be well served to consider what types of outcomes are meaningful and approximately what size the N's would have to be to show this outcome to be statistically significant.

Proportion of Variance Accounted For

There is a different way of dealing with the problem of large N's creating statistical significance in data where the magnitudes of the comparisons are too small to be of practical significance. This technique is appropriate for the F tests involved in a between-subjects ANOVA having equal N's per cell. Similar techniques for repeated-measures designs can be found in Vaughn and Corballis (1969). This technique is called "determining the proportion of variance accounted for" (for some comparison of interest). Two steps are involved: (a) performing the usual F tests involved in the experimental design, and (b) following up on any significant F or F's with a measure reflecting the size of the effect (e.g., the difference between row means) independent of sample size (N). With this technique, effect size is determined by taking a ratio reflecting the variability associated with SS_{effect} over the total variability (SS_{total}). We will present two techniques for determining the proportion of variance accounted for. The first η^2 (pronounced "āata" squared) is appropriate only for samples where inferences to populations are not made. For this, and other more technical reasons, η^2 is rarely reported in journals. We will present it here because it is easy to understand, and this understanding can be directly applied to the more complicatd measure, ω^2. This second measure, ω^2 (pronounced "omāga" squared), is widely used to estimate the effect size involving a comparison within a population. This estimate involves the ANOVA components coming from an experiment.

η^2

Recall that in an ANOVA, the total variance SS_T is broken down into components. In the one-factor ANOVA, $SS_T = SS_B + SS_W$. In other words, the total variability is broken down into that due to differences between the means of conditions (i.e., SS_B) and that due to "error" (i.e., SS_W). η^2 is defined as

$$\eta^2 = \frac{SS_{effect}}{SS_T} \qquad 12.2$$

The only effect in a one-factor ANOVA is SS_B; therefore, for a one-factor ANOVA, $\eta^2 = SS_B/SS_T$. However, a two-factor ANOVA has three different effects: SS_r, SS_{col}, and SS_I. Therefore, for a two-factor ANOVA, a value for η^2 can be computed for rows, columns, and interaction. As can be seen from the formula, η^2 is simply the ratio of the variability of the effect (which reflects the magnitude of the differences between means) over the total variability.

ω^2

ω^2 is a slightly modified η^2. This modification is necessary to give a more accurate estimation of population effects. ω^2 is computed as

$$\omega^2 = \frac{SS_{effect} - df_{effect}\, MS_{error}}{SS_T + MS_{error}} \qquad 12.3$$

Like η^2, ω^2 is a ratio reflecting the variability coming from the magnitude of an effect over the total variability involved in the data. We will illustrate the computations involved in determining η^2 and ω^2 by finding appropriate η^2 and ω^2 values for the ANOVA example from Chapter 8. The summary table from that example is reproduced as Table 12.1.

TABLE 12.1
Summary Table for Two-Factor ANOVA

Source of Variation	SS	df	MS	F
Rows (therapists)	12	1	12	5.33
Columns (clients)	0	1	0	0
Interaction	192	1	192	76.8
Error	18	8	2.25	
Total	222	11		

An ω^2 value should not be computed for F values that are not significant. We will demonstrate the computations by using the effect for the interaction that was significant and very large. Then η^2 and ω^2 for the interaction would be computed as follows:

$$\eta^2 = \frac{SS_{interaction}}{SS_{total}}$$

$$= \frac{192}{222} = .86$$

$$\omega^2 = \frac{SS_{interaction} - df_{interaction}\, MS_{error}}{SS_{total} + MS_{error}}$$

$$= \frac{192 - 2.25}{222 + 2.25} = .84$$

These values for η^2 and ω^2 are extremely large. This occurred because the example was made up to show a very large interaction in relation to other sources of Variance. With data from real experiments, η^2 and ω^2 commonly range from .2 to .3.

Comments on the Use of ω^2 The measure ω^2 has some attractive features. It allows the researcher to follow a significant F from an ANOVA with an additional measure that gives an indication of effect size while being independent of sample size. However, ω^2 also has some serious problems involving its interpretation. The problems involve the fact that ω^2 is not simply a function of effect size. It is a ratio of effect size over all of the variability components within an ANOVA. These variability components include variability due to error and, in multifactor experiments, variability due to main effects and interactions. What this means is that for a given effect size, ω^2 can vary greatly depending on things having nothing to do with effect size. The problem for multifactor experiments can be partly solved by substituting $SS_{effect} + SS_{error}$ for SS_T in the ω^2 formula. This eliminates the confounding coming from other main effects and interactions. However, ω^2 is still going to be directly influenced by error variability. This means that in situations where individual differences (and/or other factors that influence the error term) are large, ω^2 values are going to be much lower than in situations where error term components are small. For example, suppose an ω^2 of .20 is reported. One might imagine different scenarios where this value would be interpreted differently. Under one scenario, it would be interpreted as showing a large effect size, whereas under a different scenario it would be interpreted as showing a small effect size. This possible difference in interpretation across experiments is an undesirable characteristic of ω^2.

In summary, there may be situations where reporting ω^2 is useful. However, it does not appear to be the answer to the sample size problems that we have been discussing.

Point Estimates and Confidence Intervals

Point estimates are also called parameter estimates. Thus, the mean of a sample can be used as a point estimate of the population mean from which that sample was taken. In fact, in a typical experiment, we use the sample means from our experiment to estimate population means. With hypothesis testing procedures, we then test to see if we can be confident that our sample means (or differences between means) *did not* come from the population of scores that would occur if H_0 were true. With respect to estimating effect size, we can utilize the fact that the best estimate of a population mean is a sample mean from that population. Thus, the best estimate of the magnitude of the differences between population means (i.e., the effect size) is the magnitude of the differences between sample means. Further, this estimate of effect size will become more accurate as sample size increases. Therefore, with point estimates, sample sizes can never be too large.

A simple example will illustrate effect size using point estimates. Suppose that we wish to compare the effects of learning common nouns by imagery techniques versus repetition techniques. Following

point:

a parameter such as the mean of a population

a proper experiment, we find that the mean number recalled for the repetition group is 20, whereas the mean number recalled for the imagery group is 30. We therefore estimate that the imagery instructions produced a 50% increase in the number of nouns recalled compared with the repetition instructions.

Comments on Determining Effect Size by Using Point Estimates   In those rare situations where sample sizes in the thousands can be obtained, sample means will be so close to population means that statistical tests of inference will usually not be necessary. In these situations, inferring effect size by comparing sample means will be quite accurate. However, in the typical situation, involving 15 to 30 subjects per condition, considerable error may be involved in estimating effect size from comparing sample means. Recall the formula for the estimated standard deviation of means from Chapter 5: $S_{\bar{X}} = S/\sqrt{N}$. A sample mean will vary depending on two things: (a) the number of scores (i.e., N) constituting the mean, and (b) the variability within the sample. Thus, in situations where N is fairly small and S is fairly large, $\bar{X}$ can vary widely. In this situation, the sample mean, though the best single predictor of the population mean, is nevertheless not a very good predictor. *Therefore, the major disadvantage of using point estimates to determine effect size is that in certain situations the magnitude of potential error of this estimate may be rather large.*

There is a technique available for determining the magnitude of likely error involved in estimating population means from sample means. Recall from Chapter 5 that an estimate of the standard deviation of a population of sample means of size N (i.e., $S_{\bar{X}}$) is

$$S_{\bar{X}} = \frac{S}{\sqrt{N}} \qquad 12.4$$

In other words, suppose we were to repeat one of the conditions in an experiment thousands of times. Then the means from the scores of each repetition would form a set of scores that would have a particular standard deviation. An estimate of the population standard deviation of this hypothetical set of means can be obtained from Equation 12.4. We also know from Chapter 6 that when we make estimates of σ using S, the t distribution is the appropriate model from which to make inferences. Since we now know something about the range over which sample means are likely to vary based on information contained in our sample, we can make an estimate of the magnitude of error that is likely to be made when the sample mean is used as an estimate of a population mean. We do this with a procedure whereby *confidence intervals* (C.I.) are determined. *A **confidence interval** is an interval wherein a population parameter (such as the mean) will fall a certain percentage of the time.* The typical percentage used is 95% ($1 - \alpha$, where $\alpha = .05$). Thus, a 95% confidence interval will contain the specified popu-

confidence interval:

an interval within which a population parameter (such as the mean) will fall a certain percentage of the time lation parameter 95% of the time. An equivalent way of stating this is that one can be 95% confident that the specified population parameter will fall within a specified interval (thus, the term *confidence interval*).

The formula for a C.I. for μ (the population mean) where σ is estimated from S is

$$\text{C.I. } (1 - \alpha) \text{ for } \mu = \bar{X} \pm (t_{\text{df}}, \alpha)S_{\bar{X}} \qquad\qquad 12.5$$

where $\alpha = \alpha$ used in finding the critical t
          df = df in finding the critical t
          $S_{\bar{X}}$ = estimated population standard deviation of means, given by

$$S_{\bar{X}} = \frac{S}{\sqrt{N}}$$

Let us now examine the components of Equation 12.5 and see why they are appropriate for determining a confidence interval that will contain a population mean. First, as previously pointed out, the term $S_{\bar{X}}$ is the estimated population standard deviation of sample means that would occur if we kept replicating the conditions under which the sample mean of interest (such as a mean from a condition in an experiment) occurred. The term (crit. t_{df}, α) is the critical t for a given df at a chosen α level. For 95% C.I., the α value would be .05. The term (crit.t_{df}, α) tells us how many standard deviations we have to go in each direction from the center of the distribution of sample means to include 95% of these sample means. Thus, multiplying (crit.t_{df}, α) times $S_{\bar{X}}$ gives an interval within which 95% (using α = .05) of sample means would fall. Now, if our sample means were taken from a population having a known mean μ, then the expected mean of our hypothetical distribution of sample means would have the same value as μ. Also, there would be a 95% chance (using α = .05) that a particular sample mean $\bar{X}$ would fall within the interval $\mu \pm$ (crit. $t_{\text{df}}, \alpha)S_{\bar{X}}$. Notice that this model also implies the reverse, namely, that this same interval (crit.$t_{\text{df}}, \alpha)(S_{\bar{X}})$ placed around each sample mean ($\bar{X}$) in the distribution will contain the population mean μ 95% of the time as shown in Figure 12.1. Thus, given a particular sample mean, Equation 12.5 will create an interval that contains μ 95% (using α = .05) of the time.

Example What is the 95% C.I. for μ based on the following data: 22, 20, 18, 18, 22?

Step 1

$$\bar{X} = \frac{\Sigma X}{N} = 20$$

where N equals the number of scores.

Figure 12.1
Illustration of 95% confidence intervals using 4 df, $\alpha = .05$, and $S_{\overline{X}} = 1$.

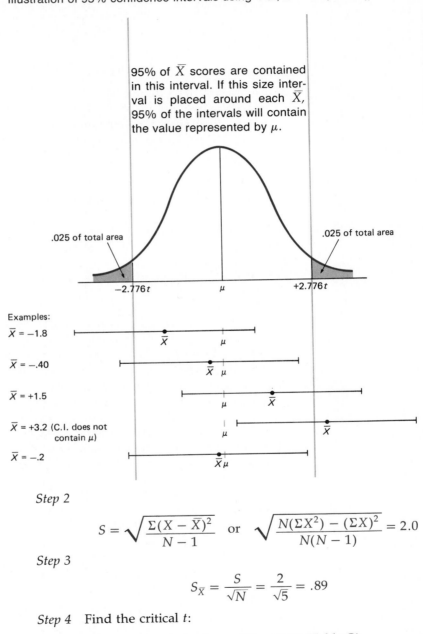

95% of $\overline{X}$ scores are contained in this interval. If this size interval is placed around each $\overline{X}$, 95% of the intervals will contain the value represented by μ.

.025 of total area   .025 of total area

$-2.776t$ μ $+2.776t$

Examples:

$\overline{X} = -1.8$

$\overline{X} = -.40$

$\overline{X} = +1.5$

$\overline{X} = +3.2$ (C.I. does not contain μ)

$\overline{X} = -.2$

Step 2

$$S = \sqrt{\frac{\Sigma(X - \overline{X})^2}{N - 1}} \quad \text{or} \quad \sqrt{\frac{N(\Sigma X^2) - (\Sigma X)^2}{N(N - 1)}} = 2.0$$

Step 3

$$S_{\overline{X}} = \frac{S}{\sqrt{N}} = \frac{2}{\sqrt{5}} = .89$$

Step 4   Find the critical t:

$$\text{crit.}t_{4\text{df}}(\alpha = .05) = 2.77 \,(\text{Table C})$$

$$\text{df} = N - 1 = 5 - 1 = 4$$

(*Note:* **A nondirectional value for *t* is always used.**)

$$\alpha = 1 - \% \text{ confidence}$$

$$= 1 - .95 = .05$$

Step 5 Thus,

$$95\% \text{ C.I. for } \mu = 20 \pm (.89)(2.7)$$

$$= 20 \pm 2.47$$

Thus, the 95% C.I. for μ ranges from 17.53 to 22.47.

In summary, the problem of very large sample sizes creating statistically significant results can be solved if the researcher looks at two different aspects of the data analysis. The first aspect involves determining whether the means are reliably different (i.e., did the statistical test show significance?). The second aspect involves evaluating whether the magnitude of the difference between the means is meaningful. It is quite acceptable to conclude that though a certain comparison is statistically significant, it is also trivial in size and has no meaningful interpretation. However, one should never interpret a comparison as meaningful when the statistical test on that comparison was not significant.

DIRECTIONAL VERSUS NONDIRECTIONAL TESTS

As we pointed out in Chapter 4, an outcome can occur in either of two directions. For example, if one is comparing learning technique B with learning technique A, it is possible for technique B to produce superior learning compared with technique A. It is also possible for technique B to produce inferior learning compared with technique A. Hypothesis testing procedures are designed to allow a researcher to be highly confident that some nonrandom event has taken place. When comparing technique B to technique A, nonrandomness can occur by B being either superior or inferior to A. For a nondirectional t test with the usual α level of .05, half of the rejection region for H_0 will be in one tail of the t distribution while the other half of the rejection region will be in the other tail. Thus, with a nondirectional test, it is possible to reject H_0 with either the effects of A > B or A < B.

Suppose a situation exists where a researcher is only interested in whether or not an outcome occurs in one direction. If all of the rejection region is placed within one tail of the t distribution, H_0 will be more easily rejected because the critical t will be smaller. This raises a long-debated issue of the appropriateness of directional tests. The crux of the issue is this. Is it appropriate to allow H_0 to be more easily rejected if the outcome of an experiment is in the direction of interest while all contrary outcomes, however extreme, are ignored? In this author's opinion, the answer to this question is yes and no (and, yes, this is a serious answer). The answer is yes if the subject area under investigation is applied only. The answer is no if the subject area under investigation is theoretical. The difference between applied (in the pure sense) and theoretical research is that with applied research, there is no attempt to ask why outcomes occur as they do, whereas with theoretical research questions involving "why" are asked. The following example will illustrate this distinction. Suppose that a teacher is using a set of programmed materials to teach a topic. Now suppose that he receives another set of programmed materials on the same topic that the publisher claims is superior to the materials that he is currently using. Suppose further that the teacher is able to randomly assign students to each set of materials and to compare sets of materials by giving a common test. This is an example of applied research. The teacher in this case is not asking why or what makes one set of materials superior; he only wants to know if the new set is better. If the new set creates either the same performance or inferior performance, he will continue using the old materials. It therefore seems appropriate for him to use a directional test. Only if the new set is superior will he reject H_0.

Let us now redo this hypothetical experiment within the context of theoretical research. In this case, the researcher not only wants to know which set of materials is best, but she wants to know why that set is best. Thus, while she may believe and/or predict that the new set of materials will create superior performance, an outcome in the opposite direction would also provide information of importance. In other words, when one is trying to understand a process, no outcome can be ignored. Therefore, when one is doing theoretical research, that is, doing research with the goal of understanding a phenomenon, then directional tests should not be used. By contrast, with applied research, directional tests may or may not be used, depending on the interest of the investigator.

There is also a more pragmatic reason for using nondirectional tests. Since a nondirectional test is more conservative with respect to rejecting H_0, the use of a nondirectional test will rarely be questioned. On the other hand, the use of a directional test may well be questioned. In this situation, the burden of proof is on the researcher to justify the use of the directional test. Thus, as a researcher, one can simplify one's life by always using the noncontroversial nondirectional test.

In the actual use of a statistical test, the only time that different conclusions will be drawn from the directional and the nondirectional tests are those rare occurrences wherein the computed t (or other statistic) value falls in the narrow range of values between the cutoff points for the two types of tests. Rather than use a controversial decision criterion, a better strategy is to try to design the experiment to be sensitive enough to show statistical significance with the more conservative nondirectional test. The usual way of doing this is to increase the sample size of the conditions in the experiment.

In summary, though there are some situations in applied research where directional tests may be appropriate, controversy can be avoided if nondirectional tests are always used with consideration given to using larger sample sizes. (See the discussion in the previous section on sample size issues.)

CHOOSING AND USING α LEVELS

According to the standard statistical model presented in this book, an α level, chosen prior to data collection, serves as a cutoff for retaining or rejecting H_0. If the probability that a set of data would occur by random factors is greater than the chosen α level, then H_0 is retained. If the probability is less than the chosen α level, then H_0 is rejected. However, if one reads journal articles that use inferential statistics (the t test, F test, χ^2 test, etc.), a curious finding emerges. One can find hundreds of examples where H_0 is both rejected and retained at $\alpha = .05$. One can also find hundreds of examples where H_0 is rejected at the $\alpha = .01$ and the $\alpha = .001$ level. However, one cannot find a single case where H_0 has been retained at the $\alpha = .01$ or the $\alpha = .001$ level. Clearly, there is a mismatch between the standard statistical model and the way that the results of statistical tests are often reported. In fact, what is often used (and condoned by most journals) is a more complicated version of the decision process involving the use of α levels. With this version, H_0 is never rejected when the test statistic (e.g., computed t or F value) shows a probability of occurring by chance that is greater than .05. However, if this test statistic shows a probability of chance occurrence that is less than .05, then the test is reported as significant at the lowest convenient level possible. Convenient levels are determined by the points marked off by standard tables such as Tables C and D in the appendix. Common points delineating areas in the tails of distributions are .05, .025, .01, .02, .005, .001, .0001. Thus, an α level of .05 is still used to retain or reject H_0. However, if H_0 is rejected, additional information is given concerning the probability of a chance occurrence of the test statistic.

The following represents a typical way of reporting a significant F from a F test. "An analysis of variance showed that this comparison was significant, $F(1,69) = 12.76$, $p < .001$." The p value represents the smallest convenient cutoff point beyond which the test statistic fell.

Currently, many journals will accept either method of reporting the results of statistical tests. One can state that $\alpha = .05$ for all reported statistical tests, and then omit p values. Or one can use $\alpha = .05$ to retain or reject H_0, and also report p values as described. However, both methods require the author to report the relevant degrees of freedom and values of the test statistics (e.g., $F(1,69) = 12.76$). From this information, the interested researcher can compute p values if he or she so desires. Therefore, the difference in reporting is a matter of preference rather than substance.

REFERENCES

Vaughn, G. M., and Corballis, M. C. (1969). Beyond tests of significance: Estimating strength of effect in selected ANOVA designs. *Psychological Bulletin, 17,* 204–213.

Welkowitz, J., Ewen, R. B., and Cohen, J. (1982). *Introductory statistics for the behavioral sciences* (3rd ed.). New York: Academic Press.

SUMMARY

Assumptions for t and F tests to be valid in a strict sense are that the populations from which samples are taken are normally distributed and have equal variances. However, these tests are robust and will give accurate results when these assumptions are moderately violated, particularly with sample sizes that are equal and above the 20–30 range.

Experimental findings can be statistically significant but have no practical significance because of small differences between the means of conditions. This situation usually occurs with very large sample sizes. There are three types of solutions to the problem of a mismatch between statistical significance and practical significance. They are

(a) Power analysis: With this technique statistical significance is equated with practical significance by first determining a meaningful effect size and estimating the variability that would be involved in the data. From this information, a sample size estimate can be obtained such that results that are statistically significant will also be of practical significance.

(b) Percent of variance accounted for: This technique, using η^2 or ω^2, gives a relative measure of effect size that can be obtained following an F test. These tests are independent of sample size, but are influenced by error variability and, in multifactor designs, by other main effect and interaction components.

(c) Point estimates: This technique uses differences in sample means to estimate corresponding differences in population means. With this technique, the magnitude of the error of estimation gets smaller as sample size gets larger. The magnitude of potential error in estimating population means from sample means can be found by determining confidence intervals. A confidence interval is an interval within which a population parameter (such as the mean) will fall a certain percentage of the time.

When an investigator is interested in understanding a process (i.e., doing theoretical research), it is recommended that nondirectional tests always be used. In applied research situations, either directional or nondirectional tests may be used, depending on the situation.

Journals generally accept two formats for reporting the results of statistical tests. One format is to use an α level of .05 to retain or reject H_0 for all tests. The other format is to use an α level of .05 to retain H_0, but, if H_0 is rejected, to report the α level at the lowest convenient level that is possible.

KEY TERMS

theoretical sampling distribution

power analysis

statistical versus practical
  significance

percent variance accounted for

η^2

ω^2

point estimate

confidence interval

directional test

nondirectional test

theoretical research

applied research

α level

PROBLEMS

1. Use the technique, similar to power analysis, described in the chapter to determine the sample size (N) that should be used in an experiment where doing a t test on the data would be appropriate.

*a   estimated S_1^2 and $S_2^2 = 40$; meaningful effect size (i.e., $\bar{X}_1 - \bar{X}_2$) = 5.

  b   estimated S_1^2 and $S_2^2 = 80$; meaningful effect size = 5.

  c   estimated S_1^2 and $S_2^2 = 40$; meaningful effect size = 10.

    d   estimated S_1^2 and $S_2^2 = 20$; meaningful effect size = 4.

    e   estimated S_1^2 and $S_2^2 = 80$; meaningful effect size = 6.

    f   estimated S_1^2 and $S_2^2 = 9$; meaningful effect size = 2.

2. Find η^2 and ω^2 for problems 2, 4, and 6 in Chapter 7.

3. Find ω^2 for the significant F's in problems 2, 3, and 4 in Chapter 8.

4. Find the 95% C.I. for μ based on

    *a $S = 4, N = 20, \bar{X} = 10$

    b $S = 12, N = 20, \bar{X} = 10$

    c $S = 12, N = 80, \bar{X} = 10$

    d $S = 10, N = 25, \bar{X} = 3.4$

5. Find the 95% C.I. for μ from the following data:

(* a)	(b)	(* c)	(d)
17	4	36	− 1
24	0	54	3
19	9	29	− 7
15	1	17	− 9
12	7	35	2
18	6	41	10
	2	30	− 4
	7		6
	5		− 3
	4		

CHAPTER 13

The Use and Misuse of Statistics

DO STATISTICS LIE?

Obviously, statistics themselves do not lie. They are only ways of handling data so that summaries may be presented and/or inferences may be made. The real issue is the extent to which people use statistics, knowingly or unknowingly, to deceive others. We will examine this in the context of both data analysis and inferences from data analysis.

CHEATING

Researchers are people, and some people are dishonest. With money and prestige often depending on the publication of one's research, should we not expect some or perhaps a lot of data to be falsified? The same pressure to cheat may also bear upon students who wish to finish a project on time or perhaps earn a better grade. Yet the documentation of cheating by professionals is very rare; and if my experience as a teacher is typical, cheating by students by fudging data is also rare (perhaps 1 person in 30 to 50).

Part of the explanation for rarity with which cheating by creating or changing data occurs is that researchers have high ethical standards. Based on my experience, this seems to be the case. However, there are also other powerful factors at work to keep professional and student researchers honest.

First, research that is important usually gets partially or completely replicated by someone else. If a researcher's work cannot be replicated, it speaks ill of either his or her integrity or ability as a researcher. This is a powerful incentive to do things properly. Second, it is difficult for the professional and very difficult for the novice to cheat by making up data. Data can be detected as erroneous because:

(a) It is beyond an appropriate range. For example, a study reporting human reaction times below 200 milliseconds has a serious problem because humans do not respond that fast. Reaction times above 500 milliseconds might also be suspect, depending on the circumstances under which they were obtained.

(b) The variability is inappropriate. Researchers within an area know approximately the magnitude of variability to expect from a set of data from an experiment related to that area. Measures of variability that are considerably smaller or larger than what would ordinarily be expected become suspect. This is particularly true if there is no explanation from the researcher for this occurrence.

(c) The data indicate nonrandomness, when randomness is to be expected. Suppose a researcher reports a series of 10 means, all of which end in the number 5, under conditions where all 10 digits are equally likely. The probability of this happening by chance is $(1/10)^{10}$ or 1 chance in 10 billion. Clearly the conclusion is that the data are erroneous.

Thus, in addition to ethics, there are other strong reasons for researchers being honest.

INAPPROPRIATE COMPARISONS

Statistics are sometimes deceptive because of the ignorance of those who are presenting them. This may show up in comparisons that make no sense. For example, I once read a statement in a popular national weekly news magazine where the median annual income of city dwellers was compared to the top 25% of dwellers in the suburbs. This, of course, makes no sense because the median and the 25th percentile reflect different parts of a distribution.

A favorite ploy of politicians is to use absolute numbers where ratios are usually used. For example, the number of unemployed in a country may be at or near an all-time high, whereas the unemployment rate may be moderate or low. This situation can occur when there has been a rapid influx of new workers.

Thus, care should be taken to compare means with means, medians with medians, standard deviations with standard deviations, rates with rates, etc.

Sometimes comparisons are inappropriate not because of explicit statements, but by implication. For example, I frequently get advertisements whereby I am guaranteed two prizes out of eight offered if I will listen to a sales pitch. Of the eight prizes, six are very expensive while two are modest in value. A drawing will determine which prizes will be won. The inference that the advertiser tries to create is that each prize is equally likely. In fact, however, if one reads the *very* fine print in the offer, the cheap prizes are several thousand times more likely to be won than the expensive prizes.

A similar problem can occur in trying to interpret statements of fact. For example, suppose you read that most automobile accidents occur within a 5-mile radius of the driver's home. Does this mean that driving near one's home is more dangerous? Not necessarily, unless it can be shown that people drive as many miles beyond a 5-mile radius of their homes as within this radius. In other words, accidents may simply be a function of the number of miles driven, and most people may drive most of their miles within a 5-mile radius of their homes.

One of the easiest ways for statistics to be misleading is to interpret them without reference to an adequate control group. For example, suppose that 80% of people who take a particular medication report relief from pain within one hour. We cannot say from these data that the medication is effective until we compare this group with a group that had taken a placebo.

"Going to college pays off financially because data show that college graduates earn more in their lifetime than non-college-graduates." This statement may be true, but the data by themselves are not convincing because of the lack of an adequate control group.

People who go to college are not selected on a random basis. The selection factors that lead people to go to college (e.g., motivation, socioeconomic factors) may contribute to their getting better jobs and earning more money.

To summarize these last examples, we may say that *without an adequate context or experimental design, statistics are meaningless.*

USING GRAPHS TO DISTORT VARIABILITY

Graphs may be used to make data *appear* to have more or less variability than is the actual case. Figure 13.1 shows two bar graphs of the same hypothetical data involving the mean number of study hours spent by freshmen, sophomores, juniors, and seniors at a particular university. From panel *a*, there appears to be little difference in study times between the classes. However, from panel *b*, there appears to be large differences between the classes. The reason that the graphs look so different is the different scales used for the *y* axis. The units in graph *b* are much longer with respect to the *y* axis than are the units in graph *a*. This gives the appearance of larger variability in graph *b* than in graph *a* although the data in both graphs are the same. The general rule that should be followed in choosing units along the *y* axis is to choose units that allow the reader to make fairly accurate estimates of the magnitude of the conditions within the graph. Thus, in our situation graph *b* is preferred. *However, care must always be taken to carefully consider the meaning represented by the chosen y axis scale.* In our case, the percent difference, between the highest and lowest bars is only about 25%, not the 200% or 300% as appears to be the case when looking at graph *b*.

Figure 13.1
Mean number of study hours per month.

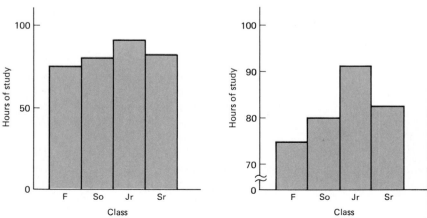

INAPPROPRIATE INFERENCES FROM AN EXPERIMENT

The usual inferences that are made in an experiment are shown in Box 13.1. The first inference is a statistical one and concerns retaining or rejecting the null hypothesis. The following represent nonstatistical inferences that occur when H_0 is rejected: The first inference concerns attributing causality to the independent variable when the null hypothesis is rejected. The second inference concerns generalizing the results beyond the specific methodology and subject population used in the experiment. There is sometimes a third inference concerning the relevance of the results to a theory or theories. This inference is beyond the scope of the current discussion. We will now examine some problems that arise in making these inferences.

BOX 13.1

Experimental Inference Chain

(The inferences can be treated independently, although it is rare for the third to be made without the first and second being true.)

Start: H_0 rejected.

First Inference: Results occurred because of I.V., not because of confounds or unknown factors. (focus: method section)

Second Inference: Results are valid beyond the specific paradigm and subject population used. (focus: sampling techniques and the topic area)

Third Inference: Results support a particular construct or theory. (focus: are there alternative explanations for the results)

Note. When H_0 is retained, then usually the preceding inferences are not made. Exceptions that take liberties with the statistics model are usually of high power.

LIMITATIONS OF H_0 RETENTION

There are many situations involving research where the experimenter would like to make a case for two or more groups being equivalent. For example, an experimenter might like to demonstrate that two groups of infants performed in equivalent fashion on a task before different training procedures were administered. Or perhaps an experimenter would like to demonstrate that learning verbal materials without awareness does not exist.

Experimenters sometimes make the mistake of using a statistical test to demonstrate that the null hypothesis is true. As we discussed in Chapter 6, when we retain H_0, we do not demonstrate, or show with any degree of confidence, that H_0 is really true. When one wants to show equivalence between means, one does so by showing that the difference between means is very small in relation to a measure of variability of the means. The situation thus involves making a reasonable case for equivalence rather than demonstrating equivalence through a statistical test.

INAPPROPRIATE CONCLUSIONS WHEN H_0 IS REJECTED

When H_0 is rejected, the implication is that some nonrandom event has occurred. It is only through the control of extraneous variables (i.e., variables other than the independent variables) that the experimenter can be confident that the independent variable/s caused the nonrandom event. When an extraneous variable(s) could also have created the outcome of an experiment, the experiment is said to be *confounded,* as illustrated in Chapter 1. How to detect confounded experiments is an important topic in its own right. However, the point to be made here is that finding statistical significance is no guarantee that the independent variable(s) has created the effect shown by the data.

confounded experiment:

an experiment that has results that can be explained by something other than the independent variable manipulation

INAPPROPRIATE GENERALIZATION TO A POPULATION

In most experiments, the results are generalized to a population that is beyond that from which random samples were taken to form the data base of the experiment. These generalizations are made on a basis other than the statistical inference model, and may or may not be justified. Justification is a matter of either logical arguments or replicating experiments in other situations. For example, suppose that a reaction time study is done on East Coast college students. One might argue that these results would generalize to all college students because both the physiology and the approach to the task would be the same for all college students. Most researchers would probably accept this argument, but the case could be strengthened if the results of the study were replicated at other colleges around the country.

The generalization problem becomes even more severe if samples taken from a population are not truly random, as is often the case in research. Because of ethical consideration, selected individuals from a population (e.g., students taking introductory psychology) cannot be forced to participate in a study. Therefore, some slippage in the random selection process is almost inevitable. The magnitude of the problem varies widely and should be evaluated on a case by case basis. Random and nonrandom samples may not vary much when selection is for routine tasks, but may vary widely when selection is for demanding tasks or tasks involving pain or embarrassment.

A famous controversy over the generalization of results occurred when the Food and Drug Administration (FDA) wanted to ban saccharin because heavy doses of saccharin were found to cause cancer in laboratory rats. The FDA generalized these results to humans and inferred a danger. Others felt this generalization to be unwarranted and thought saccharin should not be banned. Because neither side could make an overwhelming case for their position, a compromise was reached whereby saccharin was allowed to be sold and used in products, provided warning labels were attached to the product. These types of generalization problems are common when nonhumans must be substituted for humans in research designed to be applied to humans. Can you think of other controversies of this type?

CAUSATION BASED ON CORRELATION

We have already discussed in Chapter 10 the fact that a high correlation does not imply causation. If this is true, then why would a researcher bother doing a correlational study? The answer is that a correlational study *can* provide some valuable information about causation. For example, a zero correlation based on the populations of two variables indicates that a causal connection between two variables cannot exist. If a nonzero correlation is obtained, this indicates only that a causal relation may exist.

Unfortunately, some of the major issues in our society involve data that are correlational in nature. For example, a significant correlation between sex offender and the reading of pornographic materials indicates that a causal connection *may* exist. It does not indicate that there *is* a causal relationship. This must be established by other means, such as a well-controlled experimental design. Failure to recognize the appropriate relationship between correlation and causation has fueled many controversies over social issues, as well as scientific issues.

STATISTICS AND EXPERIMENTAL DESIGN

The design of an experiment requires that decisions be made that will affect the likelihood of the results showing statistical significance. We will examine some of these issues and their potential effect on a statistical test.

DEPENDENT VARIABLE ISSUES

Once an experimenter chooses a subject area, a dependent variable must be chosen that will reflect the subject area. For example, if the subject area is *helping,* how and under what circumstances will it be measured? Two criteria must be met if the dependent variable is to be sensitive to experimental manipulations.

(a) *There must be an appropriate range of potential dependent variable values.* Basically, this means that the task should not be too easy or too difficult. Helping should not be defined as the number of hours volunteered to drive a sports car, nor should helping be defined as the number of hours volunteered to carry cement bags. In either case, the data will not accurately measure the concept of interest. Let us look at a second example. Suppose we are measuring the effects of spelling techniques on spelling accuracy. If the spelling test that we give to measure performance is too easy, most people will spell all of the words correctly regardless of any effects from the spelling techniques. If the test is too hard, few may spell any words correctly. Thus an appropriate difficulty range must be chosen; otherwise, the means of the experimental conditions will be close together (which would increase the probability of nonsignificance), not because the independent variable was ineffective, but because the choice of a dependent variable measure did not allow effects to show up.

(b) *A second concern with respect to many dependent measures is their reliability* (i.e., getting the same result again under the same circumstances). For example, if we are doing a reaction-time experiment and our reaction-time measuring instrument is often inaccurate, our data will contain these inaccuracies. The usual effect of data obtained through unreliable equipment, inconsistent instructions to subjects, random distractions, and so on, is to increase the variability within the data. As discussed in Chapter 6, this increase in variability will decrease the probability of achieving statistical significance.

Figure 13.2
Hypothetical relationship between caffeine and performance.

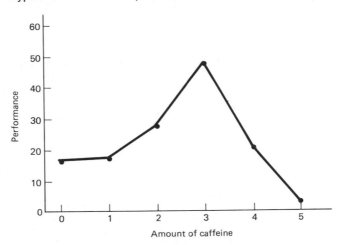

INDEPENDENT VARIABLE ISSUES

As pointed out in Chapter 1, some independent variables can be represented by different quantities. For example, suppose that we are studying the effects of caffeine on a motor performance task. Suppose further that the effects of caffeine on performance in a population is represented by Figure 13.2, where caffeine is measured by the number of caffeine pills ingested. Each pill would have a given amount of caffeine.

Now if we have only two conditions in our experiment, and we choose zero pills and one pill to represent these conditions, our statistical test will likely show nonsignificance because the population means (as shown by the zero and one pill values on the graph) are very similar in value. We would also likely get nonsignificance if we chose zero and four pills to represent our two conditions. By contrast, a zero-versus-three-pill comparison would likely show a positive effect for caffeine, whereas a zero-versus-five-pill comparison would likely show a negative effect for caffeine. Thus the results of a statistical test must be evaluated with respect to the particular choice of conditions that represent the independent variable.

From an experimental design standpoint, the way to avoid different statistical outcomes from different comparisons that may be made along a continuous dimension is to use more conditions throughout the range of the independent variable. In the current example, a detailed picture of the effect of caffeine on performance could have been obtained if six conditions had been used, one condition for each quantity (in whole numbers) of pills. Had six conditions been used, the experimenter might well have analyzed the data with a one-factor ANOVA and, if significance were obtained, followed that by protected *t* tests.

The type of function shown in Figure 13.2 is unusual. It is more common to see relationships that are fairly linear or at least monotonic (i.e., the line in the graph never changes direction with respect to the *y* axis). In designing experiments, the following guidelines should be used. When the independent variable can be represented by various quantities, conditions representing the end points should be included in the design along with one or more conditions representing the middle of the independent variable range. If a constraint exists, such as a small pool of available subjects, then conditions representing just the end points may be used.

The number of independent variables to be used in a study depends, of course, on the type of question being asked. Generally speaking, more than one independent variable is used in a design if there is a specific interest in the *interaction* between two or more variables (see Chapter 8). Following the choice of the number of independent variables comes the choice of whether to use repeated measures within the conditions of the independent variable or variables. As we discussed in Chapter 6, repeated-measures designs are generally more powerful than between-subjects designs. However, repeated-measures designs should not be used where large practice or carryover effects may exist.

THE STUDENT RESEARCH PROJECT

All research designs originate with a question that the researcher is interested in pursuing. The major difference between the way an undergraduate would typically design an experiment and the way a professor might design an experiment is that the undergraduate is usually faced with many more constraints. The most important of these constraints are time to complete the project and resources with which to carry out the project. Although these constraints will eliminate many research areas, many others exist that are both viable and important.

CHOOSING THE AREA OF THE DEPENDENT VARIABLE

The choice of a research area will be an important factor in the success of a student research project. The trick is to find a viable research area that will excite and challenge you. The many hours necessary to read in the chosen area and to carry out the project will then become fun instead of work. My students have always rated the individual research project experience very highly. Let us now look at some constraints that the student must work within in choosing the subject area for the research.

Ethics
Your project must be approved by an ethics board. Your instructor will give you information about how this works at your college or university. Although you may get routine approval for your project, do not expect it. It is therefore prudent to get an early start so that if you must change your design or project area, you will have time to do it.

Materials and Equipment
Be sure that you know exactly what materials and equipment you will need. Then be sure that it will be available when you need it. For example, if you intend to use a standardized paper-and-pencil test, is it currently accessible or will you have to send away for it? If you need norms, are they available? If you have to make up your own dependent variable measure, can you do it? For example, if you are comparing the smelling capabilities of smokers versus nonsmokers,

can you develop an adequate technique for measuring smelling capabilities?

Subject Availability

Who will your subjects be? How many will you need? Where will you get them? These three questions must have realistic answers before the research project can get under way. All experiments need a minimum number of subjects to have adequate power. This number will typically vary between 10 and 20 subjects per cell in the experimental design. The availability of subjects along with time constraints will determine how extensive your experimental design can be and still have the power to find statistical significance for differences that may exist between conditions.

Time

Most student research projects have deadlines that make extensive studies impossible. The most common mistake that students make in doing individual projects is not to allow themselves adequate time to write the report of the project. In designing your experiment, you need to estimate how long it will take to ''run'' one subject in your experiment. Then estimate the minimum number of subjects that you will need per condition in your experiment. Estimating the total time involvement per condition will allow you to determine how extensive your design can be and still allow you to complete the data-gathering phase on schedule.

THE EXPERIMENTAL DESIGN

The student should develop a tentative experimental design in the very early stages of planning a project. For example, suppose one is interested in studying the effects of noise intensity and task difficulty on the performance of puzzle-solving tasks. One would first decide the nature of the task (i.e., operationalize the task). Let's say we chose solving anagrams. One would then draw a tentative experimental design, such as using three levels of noise in combination with three levels of anagram difficulty. We might measure noise with a meter or by measuring the distance from the noise at which the noise can just be heard above a quiet background by a person whose hearing has been tested and found to be normal. We might vary anagram difficulty by using five-, six-, or seven-letter anagrams. Our tentative experimental design would then be a 3 x 3 factorial as illustrated on p. 300.

We would then decide whether a subject could participate in only one condition or could participate in all of the conditions of each independent variable without confounding carryover effects being present. In this situation, there may be concern about practice effects coming from solving several puzzles and from fatigue effects if a subject participates in many conditions. We may then decide to make task

Task Difficulty

	Easy	Medium	Hard
High			
Noise Level Medium			
Low			

difficulty a between-subjects variable and noise level a within-subjects variable. We now have a mixed design (see Chapter 9).

Suppose we now believe that we can count on getting 20 subjects for a maximum of one hour of participation apiece. We might then decide to have each subject solve three anagrams in each condition. (Ideally, we would pretest the anagrams to verify that the difficulty level was what we were expecting.) Our dependent measure would then be the total time that a subject used to solve three anagrams for each condition. Using three anagrams instead of one should remove some variability due to different levels of difficulty of anagrams within each of the three conditions.

Because we only have 20 subjects available, we may decide to reduce our design from three to two levels (easy and hard) of task difficulty. This will allow us to have 10 subjects per cell instead of 6 or 7, which will allow us to have a more powerful design. The cost is the elimination of the medium level of task difficulty, which one might speculate not to be critical to the overall hypothesis. Thus, the 3 x 2 mixed design is our final choice.

As is always the case, we had to make several subjective decisions along the way. One usually makes these decisions either from experience or from reading and copying from similar procedures as found in the literature on the topic under investigation. Good judgment also helps.

THE LITERATURE SEARCH

The literature search should go hand in hand with the development of the experimental design. You might start with a general question such as the effects of noise on performance, and then refine the question after reading some literature on the topic. Or you may develop the design after reading several articles on the topic. In any case, your experimental design ought to be systematically related to a set of studies that you will cite in the introduction section of your report. How extensive your literature search should be depends on your instructor. Your instructor is also in the best position to guide you to relevant resources. However, generally, a journal of abstracts such as *Psycho-*

logical Abstracts and references in chapters from books and recent journal articles are common sources.

AFTER THE EXPERIMENT: PRESENTING THE DATA

A reader of your report will want to know three things:

(a)   *What does a summary of your data look like?* You will need to show the reader summary statistics. Typically, this summary would be in the form of a table showing the means and standard deviations of the conditions in the experiment. For experiments having many conditions, a graph may also be an appropriate way of showing the means from the conditions.

(b)   *What are the results of the statistical analysis?* After the data are presented in summary form, you will need to perform and report the appropriate statistical tests. The analyses appropriate for most student research projects are presented in this text. An appropriate format for reporting the results of the statistical analyses is also presented in the appropriate chapters in this text.

(c)   *What is your interpretation of the data in the light of the statistical tests?* For comparisons that were statistically significant, you generally will need to discuss why this was the case. With a tight experimental design, this amounts to discussing why the independent variable or variables were effective.

Nonsignificant comparisons may be due to any or all of three reasons:

(a)   The independent variable or variables were ineffective.
(b)   Some extraneous variable or variables were masking the effect of the independent variable. For example, the task may have been too easy or too difficult.
(c)   The experimental design may not have been powerful enough to show the effect from the independent variable. Lack of power may be due to too few subjects or high variability of the data within the conditions of the experiment.

As you can see, you can make much more definitive arguments about what happened in your experiment if the results of a statistical test are significant as opposed to being nonsignificant.

THE WRITTEN REPORT

The format used for publishing journal articles is typically used for student reports. This format is formally presented in the *Publication Manual of the American Psychological Association.* This means that the report will contain five sections plus a listing of references at the end of the paper. The first section is the *abstract.* This is typically about a

150-word summary of the paper. The second section is the *introduction*. Here is where background relevant to your research is given. The context that you establish in your introduction should show a logical relationship to your experimental design. The third section contains the *method*. This section usually consists of at least three parts. The first part tells the source of your subjects and the criteria for selecting them. The second part describes the materials used in the study. The third part describes the procedure used in the study in enough detail so that it would be possible to replicate the study. When experimental designs are elaborate, there may also be a fourth section describing the experimental design.

The fourth section of the paper is the *results*. In this section, you will show and describe your tables and graphs and present your statistical analysis. The fifth section is the *discussion*. Here you analyze your results and relate them to the context that you established in the introduction. A critique of your study may also be done in this section.

One final suggestion. A short, well-written, published article from which you can pattern your sections will be of considerable help.

BOX 13.2
Checklist for Student Research Project

_____ Has project been approved by instructor or review board?

_____ Do you have a source for obtaining subjects?

_____ Do you have all necessary materials and equipment?

_____ Do you have your experimental design finalized?

_____ Have you written out a complete set of instructions?

_____ Do you have a location to run your experiment?

_____ Do you have approved consent forms and debriefing sheets?

_____ Do you know how the dependent variable is to be measured?

_____ Have you run at least two practice subjects to confirm the viability of your procedures?

_____ Do you have a way of marking your data with respect to the condition it represents?

SUMMARY

Statistics are sometimes used, knowingly or unknowingly, to misrepresent a situation. Some of the ways that this is done are the following: (a) Cheating: This is a rare occurrence among professionals, most likely because of high ethics among researchers. However, the fact that research may be replicated by someone else and the fact that cheating without getting caught is difficult, are both incentives to keep researchers honest. (b) Inappropriate comparisons: For a comparison to be meaningful, the identical descriptive measure must be used. For example, a mean must be compared with a mean, not a median. Statistics also need to be interpreted within an appropriate context and with appropriate control groups. (c) Graphical distortion: Graphs can be drawn to make data *appear* to have more or less variability than they actually possess. This is done by making the y units on the graph small or large.

Sometimes mistaken conclusions are drawn based on inappropriate inferences from an experiment. Some of these mistakes are the following: (a) Stating with confidence that H_0 is true when H_0 is retained following a statistical test. The statistical inference model only allows one to reject H_0 with a certain degree of confidence. (b) Automatically concluding that the independent variable was effective when H_0 is rejected. This conclusion is warranted only to the extent that no confounding variables are present. (c) Making inappropriate generalizations to a population: The statistical model allows one to generalize the results of an experiment only to the population from which random samples were taken. Generalizing beyond this population is done by replicating the experiment with other populations and by making logical arguments. (d) Inferring causation from correlation: When a zero correlation exists between two variables, one can infer that the two variables are not causally related. However, if a nonzero correlation, even a high one, does exist, this indicates only that a causal relationship *may* exist between the two variables.

Decisions involved in the design of an experiment will affect the likelihood of that experiment showing statistical significance. With respect to the dependent variable, it must be measured in a way that will be sensitive to the experimental manipulations. When one is choosing conditions to represent an independent variable, that choice can affect the statistical tests and the conclusion drawn from the experiment. When the independent variable can be represented by different quantities, conditions representing a high and low quantity should be included in the design. If resources permit, conditions representing middle quantities should also be included.

The student research project is constructed in the same fashion as professional research projects. The major difference is in the increased number of constraints faced by the student, particularly in time, materials, and subject resources. Ordinarily, the first step is to choose a dependent variable in an area that is both practical and of interest to the student. A tentative experimental design is then constructed, which may be modified with increased familiarity with the literature in the area or through pilot work. Typically the project is written in five sections with a listing of references at the end. The first section is the *abstract*. This is about a 150-word summary of the project. Next is the *introduction*, where literature relevant to the experimental design is discussed. The third is the *method* section, where the procedures used in the experiment are described in enough detail that another researcher could replicate the study. The fourth section contains the *results*, wherein the tables, graphs, and statistical analyses are presented. The last section is the *discussion*, where the results are critiqued and discussed within the context set forth in the *introduction*.

Appendix A

Math Review

Common Symbols

Symbol	Meaning	Example
$=$	Equals	$3 + 2 = 4 + 1$
$\neq$	Does not equal	$3 + 2 \neq 5 + 3$
$+$	Add	$3 + 2 = 5$
$-$	Subtract	$3 - 2 = 1$
$\pm$	Add and subtract	$3 \pm 2 = 3 + 2$ and $3 - 2$
( )( )	Multiply values within parenthesis	$(4)(3) = 12$
$\times$ or $\cdot$	Multiply	$4 \times 3 = 12$ $4 \cdot 3 = 12$
/ or $\frac{(\ )}{(\ )}$	Divide	$6/2 = 3$ $\frac{6}{2} = 3$
$>$	Is greater than	$5 > 4$
$<$	Is less than	$5 < 7$
$\geqslant$	Equal to or greater than	$8 \geqslant 8,7$
$\leqslant$	Equal to or less than	$8 \leqslant 8,10$
$(\ )^2$	Square of (a value multiplied by itself)	$(3)^2 = 3 \times 3 = 9$
$\sqrt{}$	Square root of	$\sqrt{9} = 3$
$\|\ \|$	Absolute value of (negative signs are disregarded)	$\|+4\| = 4$, $\|-4\| = 4$
$\ldots$	Continue on	$2,4,6,8, \ldots$
X, Y or other letters	Represents a set of numbers	X: 1, 5, 7, 3
Σ	Add the following numbers	ΣX, where X represents 1,5,7,3 $=$ $1 + 5 + 7 + 3 = 16$

Positive and Negative Numbers Assumption: If a number has no sign, it is understood to be positive; i.e., $6 = +6$.

ADDING AND SUBTRACTING

(a)   Adding a negative number is the same as subtracting that number. *Example*: $7 + (-4) = 7 - 4 = 3$

(b)   Subtracting a negative number is the same as adding that number. *Example*: $7 - (-5) = 7 + 5 = 12$

(c)   Multiplying and dividing signed numbers: If the number of minuses in the sequence of numbers to be multiplied is odd, the result will be negative. If even, the result will be positive.

$$\begin{aligned}
\textit{Example:} \quad (-3)(-2) &= 6 \\
(+3)(-2) &= -6 \\
(-1)(+5)(+4) &= -20 \\
(-3)(-2)(-2) &= -12 \\
4 \times 3 \times 1 \times 2 &= 24 \\
(-1)(-1)(-1)(-10) &= 10 \\
(3)(2)(-2)(3) &= -36
\end{aligned}$$

Note: If properly entered, most calculators will automatically keep track of the sign.

FRACTIONS

A fraction can be turned into a decimal by dividing the numerator by the denominator.

$$\text{Example:} \quad \frac{1}{2} = .5$$

$$\frac{3}{4} = .75$$

$$\frac{4}{10} = .4$$

Thus

$$\frac{3}{4} - \frac{1}{2} = .75 - .5 = .25$$

$$\frac{1}{2} \times \frac{4}{10} = (.5)(.4) = .2$$

With calculators, it is usually easier to convert fractions into decimals and then carry out the required operations as the preceding examples illustrate. However, in some cases, it may be easier to use the following rules:

(a)   To add or subtract fractions, the denominators must be the same. All operations are then carried out in the numerator, leaving the denominator unchanged.

Example: $\dfrac{3}{4} + \dfrac{1}{4} - \dfrac{2}{4} = \dfrac{3 + 1 - 2}{4} = \dfrac{2}{4}$

$$\dfrac{2}{7} - \dfrac{1}{7} - \dfrac{4}{7} = \dfrac{2 - 1 - 4}{7} = \dfrac{-3}{7}$$

(b) To multiply fractions, multiply the numerators to get the resultant numerator and the denominators to get the resultant denominator.

Example: $\left(\dfrac{2}{10}\right)\left(\dfrac{3}{10}\right) = \dfrac{6}{100}$ or .06

$$\left(\dfrac{1}{2}\right)\left(\dfrac{3}{4}\right)\left(\dfrac{-1}{3}\right) = \dfrac{-3}{24} \text{ or } -.125$$

SQUARE ROOTS

A square root of a number is a number that when multiplied by itself will give the original number.

Example: $\sqrt{9} = 3$ because $(3)(3) = 9$

(a) The square root of a sum does not equal the sum of the square roots.

Example: $\sqrt{16 + 9} = \sqrt{25} = 5$

$$\sqrt{16} + \sqrt{9} = 7$$

(b) The square root of a product equals the product of the square roots.

Example: $\sqrt{(9)(4)} = \sqrt{36} = 6$

$$\sqrt{(9)(4)} = (\sqrt{9})(\sqrt{4}) = (3)(2) = 6$$

(c) The square root of a fraction equals the square root of the numerator divided by the square root of the denominator.

Example: $\sqrt{\dfrac{100}{25}} = \dfrac{\sqrt{100}}{\sqrt{25}} = \dfrac{10}{5} = 2$

$$\sqrt{\dfrac{100}{25}} = \sqrt{4} = 2$$

ROUNDING NUMBERS

When the first number to be dropped has a value equal to or greater than 5, increase the number to the left by one unit. Otherwise, the number to the left stays the same.

Example: (rounding to the nearest hundredth)
15.1649 rounds to 15.16
41.0497 rounds to 41.05
.51769 rounds to .52
113.9961 rounds to 114.0
5.951 rounds to 5.95

ORDER OF OPERATIONS

(a) If an expression contains only single numbers, the order for performing operations is as follows:

a. square or square root
b. multiplication or division
c. addition or subtraction

$$\text{Example: } 5 + \frac{\sqrt{9}}{3} = 5 + \frac{3}{3} = 5 + 1 = 6$$

$$\frac{10}{\sqrt{4}} - 3 = \frac{10}{2} - 3 = 5 - 3 = 2$$

$$(10)(2)^2 + 5 = (10)(4) + 5 = 40 + 5 = 45$$

(b) Expressions should be treated as single numbers when they appear in parentheses, square root signs, or in the top or bottom of fractions.

$$\text{Example: } 4(4 - 1) = (4)(3) = 12$$

$$\sqrt{30 - 5} = \sqrt{25} = 5$$

$$\frac{(6)(3)}{4 - 1} = \frac{18}{3} = 6$$

(c) When expressions are contained one within another, start from the inside and work toward the outside to find the solution.

$$\text{Example: } \frac{[(6 - 3)^2 + \sqrt{10 - 1}]}{2^2 + 2} = \frac{3^2 + \sqrt{9}}{4 + 2} = \frac{9 + 3}{6} = \frac{12}{6} = 2$$

$$\sqrt{\frac{3^2 - \frac{(4)^2}{4}}{5}} = \sqrt{\frac{9 - \frac{16}{4}}{5}} = \sqrt{\frac{9 - 4}{5}} = \sqrt{\frac{5}{5}} = \sqrt{1} = 1$$

SUMMATION

Let X represent the following numbers: 4, 1, 5, 6. The symbol Σ means add the following.

(a) If no markers are on the Σ, then add all the numbers of the designated set

Example: $\Sigma X = 4 + 1 + 5 + 6 = 16$

(b) If markers are placed on the top and bottom of Σ (i.e., $\sum_{1}^{3}$), then the bottom number designates which number in a series to start summing, and the top number designates which number in a series ends the summing. Thus $\sum_{1}^{13}$ says to start with the first number in series and sum through the thirteenth.

Example: $(X: 4, 1, 5, 6)$

$$\sum_{1}^{3} X = 4 + 1 + 5 = 10$$

$$\sum_{2}^{4} X = 1 + 5 + 6 = 12$$

(c) $\Sigma X^2 \neq (\Sigma X)^2 \quad (X: 4, 1, 5, 6)$

$$\Sigma X = 4 + 1 + 5 + 6 = 16$$

$$(\Sigma X)^2 = (16)^2 = 256$$

$$\Sigma X^2 = 4^2 + 1^2 + 5^2 + 6^2 = 16 + 1 + 25 + 36 = 78$$

(d) The following represents a very common sequence of operations in statistics representing a standard deviation (S) (see Chapter 3):

$$\sqrt{\frac{N\Sigma X^2 - (\Sigma X)^2}{(N)(N-1)}}$$

where X represents the numbers in a set and N represents the number of numbers in the set. Thus, when X represents 4, 1, 5, 4, 7, 3, $N = 6$, and $\Sigma X = 4 + 1 + 5 + 4 + 7 + 3 = 24$; $\Sigma X^2 = 16 + 1 + 25 + 16 + 49 + 9 = 116$; and

$$S = \sqrt{\frac{(6)(116) - (24)^2}{(6)(5)}}$$

$$= \sqrt{\frac{696 - 576}{30}}$$

$$= \sqrt{\frac{120}{30}}$$

$$= \sqrt{4}$$

$$= 2$$

PRACTICE PROBLEMS

1. $2 + 3 - (-1) - (-10) =$
2. $-6 + 1 + (-2) - (-3) =$
3. $(-6)(-1)(-2) =$

4. $\dfrac{(5)(4)(-1)}{4} =$

5. $\sqrt{5 + 5 - 1} =$

6. $\dfrac{(6)(5)}{7 + 3} =$

7. $\dfrac{(3 + 1)(2 + 2)}{10 - 2} =$

8. $\sqrt{\dfrac{3^2 + (2 + 1)^2 - 2}{4}} =$

9. $\dfrac{1}{6} + \dfrac{4}{6} - \dfrac{3}{6} =$

10. $\dfrac{3}{10} - \dfrac{7}{10} - \dfrac{5}{10} =$

11. $\left(\dfrac{3}{10}\right)\left(\dfrac{5}{10}\right) =$

12. $\left(\dfrac{3}{4}\right)\left(\dfrac{4}{3}\right) =$

13. $\dfrac{3}{4} + \dfrac{1}{10} =$

14. $\dfrac{9}{10} + 2.1 =$

15. $\sqrt{4}\,\sqrt{25} =$

16. Round to nearest hundredth: 4.1151; -3.4449; .997; .005001.
17. Find ΣX, ΣX^2, $(\Sigma X)^2$ when $X = 5,1,5,2,2$.
18. Find ΣX, ΣX^2, $(\Sigma X)^2$ when $X = 10,5,5,4$.
19. Find S for problem 17.
20. Find S for problem 18.

ANSWERS

1. 16
2. -4
3. -12
4. -5
5. 3
6. 3
7. 2
8. 2
9. $\dfrac{2}{6}$
10. $\dfrac{-9}{10}$

11. $\dfrac{15}{100}$
12. $\dfrac{12}{12}$
13. $\dfrac{17}{20}$
14 3.0
15. 10.
16. 4.12, -3.44, 1.00, .01
17. 15; 59; 225
18. 24; 166; 576
19. 1.87
20. 2.71

Appendix B

TABLE A
Table of Random Digits

51550	64967	31570	15748	19159	38174	51078	79811	39183	57527
96550	85168	28824	47466	56993	13151	96664	29735	70251	01079
04314	77714	11507	01440	48415	31984	99915	20282	26524	18057
04992	40521	98108	84045	91961	79256	72244	25788	05487	23595
73302	14205	08925	27625	64343	28821	37992	67156	83320	31106
10884	30735	15067	51091	15668	48777	50770	19169	76504	41165
29749	92812	08065	66782	26841	01411	95461	61134	18699	52261
60469	81373	44825	11448	73320	30151	56991	31372	06655	36472
86292	30247	30931	21029	53410	09859	37267	47514	03492	49008
94727	25234	40546	53417	36492	25723	76227	58486	15979	34876
09574	34392	03751	36933	83921	65108	63135	67572	40184	21098
95810	64584	90761	25619	57242	76482	96499	37315	81969	03466
78142	37846	90412	82889	06600	98255	09561	94876	49408	26942
73496	03542	22227	96491	25875	01152	80705	02580	48462	19399
92623	53975	07021	84038	27823	47118	36598	97426	02751	35051
34437	40331	12956	89221	33818	05023	24518	71534	67037	20882
13175	46800	69024	26844	32030	44324	44759	67653	79967	37005
04593	99638	21705	43196	78141	98290	07661	56842	05103	46286
15771	81941	46902	08846	19698	20141	19235	56149	18613	71430
88235	40413	22041	82638	31923	87851	60190	98420	83623	22256
40261	54379	31583	90636	08532	19701	27314	96290	18183	97675
91974	61993	08387	96734	96106	86097	38743	64755	16096	55722
60583	83480	44646	69643	35873	32147	66468	27892	70063	65537
81905	37983	51534	56074	68758	85313	96167	56636	60887	60487
93196	75130	30288	16102	58743	27249	30144	10204	46755	75861
49480	86917	13733	01726	38070	68028	84835	98513	38751	88801
48661	73128	68236	85198	20306	35789	60564	52648	29939	48412
06949	74058	73754	07886	50418	89923	03556	30213	87480	79956
76227	40741	09925	64011	96655	74633	80518	10995	68193	41510
34443	81242	81385	24755	52942	01798	47956	19667	44258	17309
29079	35216	35551	08988	73127	51276	18496	29077	28724	91389
87734	26492	21382	65609	00637	54091	37478	69470	02668	62034
64879	46395	92454	05241	98155	65899	25066	66441	20567	77272
13318	51221	79310	28474	53768	05223	47840	39115	79715	93025
78295	97348	48492	76936	43858	12843	01308	72080	40935	43485
21196	02168	74073	51797	41075	02065	48170	51571	06868	23706
23624	10314	44840	79228	24659	54802	77427	18487	17570	86352
24251	94281	35837	57576	65164	35511	54848	73895	70038	74998
84901	12789	57167	71992	36827	03897	77676	15448	74468	12892
68198	44962	76132	20399	33175	98806	46481	62064	18392	20337

TABLE A *(continued)*

73794	35069	66460	92803	49708	45822	11274	99466	63925	97053
99029	31669	35895	67444	98389	12308	76829	29653	42936	55884
48922	73463	27695	89902	65008	31317	96426	87228	00946	88913
08122	18694	44926	74638	22151	90254	42442	69594	94101	01665
04533	08658	23490	08019	98348	98753	29791	43662	64079	71172
21022	29925	38565	76242	36785	90287	89078	35300	53590	52922
19441	33137	70814	91340	36828	88665	97012	48621	92302	78150
59217	01488	67561	52498	51520	79799	63514	67776	88019	94351
17365	63561	54864	77101	77629	19429	20227	79902	68902	10967
58833	62286	06818	21687	92806	40254	84022	72645	30686	26409
49487	52802	28867	62058	87822	14704	18519	17889	45869	14454
29480	91539	46317	84803	86056	62812	33584	70391	77749	64906
25252	97738	23901	11106	86864	55808	22557	23214	15021	54268
02431	42193	96960	19620	29188	05863	92900	06836	13433	21709
69414	89353	70724	67893	23218	72452	03095	68333	13751	37260
77285	35179	92042	67581	67673	68374	71115	98166	43352	06414
52852	11444	71868	34534	69124	02760	06406	95234	87995	78560
98740	98054	30195	09891	18453	79464	01156	95522	06884	55073
85022	58736	12138	35146	62085	36170	25433	80787	96496	40579
17778	03840	21636	56269	08149	19001	67367	13138	02400	89515
81833	93449	57781	94621	90998	37561	59688	93299	27726	82167
63789	54958	33167	10909	40343	81023	61590	44474	39810	10305
61840	81740	60986	12498	71546	42249	13812	59902	27864	21809
42243	10153	20891	90883	15782	98167	86837	99166	92143	82441
45236	09129	53031	12260	01278	14404	40969	33419	14188	69557
40338	42477	78804	36272	72053	07958	67158	60979	79891	92409
54040	71253	88789	98203	54999	96564	00789	68879	47134	83941
49158	20908	44859	29089	76130	51442	34453	98590	37353	61137
80958	03808	83655	18415	96563	43582	82207	53322	30419	64435
07636	04876	61063	57571	69434	14965	20911	73162	33576	52839
37227	80750	08261	97048	60438	75053	05939	34414	16685	32103
99460	45915	45637	41353	35335	69087	57536	68418	10247	93253
60248	75845	37296	33783	42393	28185	31880	00241	31642	37526
95076	79089	87380	28982	97750	82221	35584	27444	85793	69755
20944	97852	26586	32796	51513	47475	48621	20067	88975	39506
30458	49207	62358	41532	30057	53017	10375	97204	98675	77634
38905	91282	79309	40922	17405	18830	09186	07629	01785	78317
96545	15638	90114	93730	13741	70177	49175	42113	21600	69625
21944	28328	00692	89164	96025	01383	50252	67044	70596	58266
36910	71928	63327	00980	32154	46006	62289	28079	03076	15619

TABLE A *(continued)*

48745	47626	28856	28382	60639	51370	70091	58261	70135	88259
32519	91993	59374	83994	59873	51217	62806	20028	26545	16820
75757	12965	29285	11481	31744	41754	24428	81819	02354	37895
07911	97756	89561	27464	25133	50026	16436	75846	83718	08533
89887	03328	76911	93168	56236	39056	67905	94933	05456	52347
30543	99488	75363	94187	32885	23887	10872	22793	26232	87356
68442	55201	33946	42495	28384	89889	50278	91985	58185	19124
22403	56698	88524	13692	55012	25343	76391	48029	72278	58586
70701	36907	51242	52083	43126	90379	60380	98513	85596	16528
69804	96122	42342	28467	79037	13218	63510	09071	52438	25840
65806	22398	19470	63653	27055	02606	43347	65384	02613	81668
43902	53070	54319	19347	59506	75440	90826	53652	92382	67623
49145	71587	14273	62440	15770	03281	58124	09533	43722	03856
47363	36295	62126	42358	20322	82000	52830	93540	13284	96496
26244	87033	90247	79131	38773	67687	45541	54976	17508	18367
72875	39496	06385	48458	30545	74383	22814	36752	10707	48774
09065	16283	61398	08288	00708	21816	39615	03102	02834	04116
68256	51225	92645	77747	33104	81206	00112	53445	04212	58476
38744	81018	41909	70458	72459	66136	97266	26490	10877	45022
44375	19619	35750	59924	82429	90288	61064	26489	87001	84273
23272	10134	04997	93413	13198	97829	32764	88678	82561	13238
58711	42535	02318	25006	56504	78133	66812	10830	89531	01124
66558	56590	26356	87273	62048	24773	52154	64570	29357	56970
54871	14219	43643	09130	89036	06995	50337	20356	29584	68668
47195	48474	14214	03616	62469	57633	76679	73178	56465	11531
60104	72179	54310	54486	80457	25760	63973	88170	95272	26897
54100	09707	85401	66598	24591	19362	99596	89509	05014	69008
36130	21395	92589	31393	59652	31274	54963	16347	45708	74308
38327	16726	34542	48252	49499	53231	12845	97958	24570	50647
97599	60943	94171	39617	01427	49491	07542	80751	29850	16570
39253	15115	04802	72920	19967	75497	24271	38957	37143	42552
08307	92336	83875	09632	43840	42836	52675	58351	29858	44523
58179	59316	94304	56221	69000	01468	72898	53386	33444	80232
87367	49333	75409	06472	13873	21792	74876	19441	25217	74254
44112	04204	58044	04701	18840	30494	92058	93703	36769	36291
56729	44393	14162	64646	98068	59893	47126	24938	38723	47892
19213	95675	85931	10915	28812	14971	32886	31398	54156	82734
22123	44315	66463	04904	24995	53484	17661	80988	00899	73514
16144	01484	06289	96375	59987	56385	95850	14377	00155	45641
42376	86111	05194	00255	08556	50857	13995	52668	53572	62478

TABLE A *(continued)*

88246	99866	99208	73413	93534	81600	62052	81537	43686	32734
73545	10949	16122	12530	10556	87512	14161	11191	90823	87042
97689	51474	18758	04684	77454	04075	14746	91078	48533	93461
11754	47497	32235	11575	63961	74084	62858	27272	92874	57587
62887	40812	11848	72001	45139	81761	42808	47885	87239	61063
35680	37748	06760	55612	16453	48694	70662	04437	55501	16065
66481	30935	21086	22973	14277	34257	01736	63187	04731	62381
49622	33691	03414	23515	14142	46561	61206	28067	78283	58398
70351	52771	62362	52775	00716	56636	68700	42714	07446	05552
63644	63401	60425	35395	68769	93943	72643	07188	50273	57643
67159	20184	50614	12026	96515	21276	37268	68944	06702	16804
37024	33605	77471	87973	95693	00793	52123	56377	36245	15931
17723	27030	58306	76595	66319	43152	39400	08434	79041	59252
17484	16024	59100	72542	06834	49385	32313	62184	03297	12909
73039	70162	02054	37426	62262	72386	12898	30680	84218	27848
62773	23503	14123	08261	66471	12315	22723	95700	71975	20950
12383	15131	49629	93222	12178	65145	47957	62964	66241	13289
38447	77456	94903	58493	85819	54765	05969	83244	82565	84463
39836	87711	06057	32420	58735	89368	11376	10812	65244	10097
01885	90793	62791	85116	56746	76105	84148	94501	41866	49672
35010	38179	97984	42756	39488	14665	23381	24963	74942	33803
77716	39886	70130	47074	53174	74156	64869	21343	38945	36492
43774	95744	42186	29122	00867	69337	53069	39416	50339	42950
44282	30459	25542	78432	88011	00796	70762	07232	42349	92736
79402	34862	67309	52596	57288	42699	10452	52772	16599	60835
72185	17542	67783	43987	00204	15691	22154	39604	79911	63971
07691	70414	72686	93011	86926	00952	50167	11320	14320	41668
20047	56581	64188	50469	83714	66723	20290	11181	80353	43619
84230	68892	42240	70066	50846	37636	58243	18411	23787	20888
93352	34119	23723	96576	10360	00012	63927	75970	85312	33417

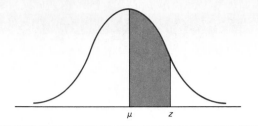

TABLE B
Proportions of Areas under the Normal Curve

(A) z	(B) area between mean and z	(A) z	(B) area between mean and z	(A) z	(B) area between mean and z
0.00	.0000	0.50	.1915	1.00	.3413
0.01	.0040	0.51	.1950	1.01	.3438
0.02	.0080	0.52	.1985	1.02	.3461
0.03	.0120	0.53	.2019	1.03	.3485
0.04	.0160	0.54	.2054	1.04	.3508
0.05	.0199	0.55	.2088	1.05	.3531
0.06	.0239	0.56	.2123	1.06	.3554
0.07	.0279	0.57	.2157	1.07	.3577
0.08	.0319	0.58	.2190	1.08	.3599
0.09	.0359	0.59	.2224	1.09	.3621
0.10	.0398	0.60	.2257	1.10	.3643
0.11	.0438	0.61	.2291	1.11	.3665
0.12	.0478	0.62	.2324	1.12	.3686
0.13	.0517	0.63	.2357	1.13	.3708
0.14	.0557	0.64	.2389	1.14	.3729
0.15	.0596	0.65	.2422	1.15	.3749
0.16	.0636	0.66	.2454	1.16	.3770
0.17	.0675	0.67	.2486	1.17	.3790
0.18	.0714	0.68	.2517	1.18	.3790
0.19	.0753	0.69	.2549	1.19	.3830
0.20	.0793	0.70	.2580	1.20	.3849
0.21	.0832	0.71	.2611	1.21	.3869
0.22	.0871	0.72	.2642	1.22	.3888
0.23	.0910	0.73	.2673	1.23	.3907
0.24	.0948	0.74	.2704	1.24	.3925
0.25	.0987	0.75	.2734	1.25	.3944
0.26	.1026	0.76	.2764	1.26	.3962
0.27	.1064	0.77	.2794	1.27	.3980
0.28	.1103	0.78	.2823	1.28	.3997
0.29	.1141	0.79	.2852	1.29	.4015
0.30	.1179	0.80	.2881	1.30	.4032
0.31	.1217	0.81	.2910	1.31	.4049
0.32	.1255	0.82	.2939	1.32	.4066
0.33	.1293	0.83	.2967	1.33	.4082
0.34	.1331	0.84	.2995	1.34	.4099
0.35	.1368	0.85	.3023	1.35	.4115
0.36	.1406	0.86	.3051	1.36	.4131
0.37	.1443	0.87	.3078	1.37	.4147
0.38	.1480	0.88	.3106	1.38	.4162
0.39	.1517	0.89	.3133	1.39	.4177
0.40	.1554	0.90	.3159	1.40	.4192
0.41	.1591	0.91	.3186	1.41	.4207
0.42	.1628	0.92	.3212	1.42	.4222
0.43	.1664	0.93	.3238	1.43	.4236
0.44	.1700	0.94	.3264	1.44	.4251
0.45	.1736	0.95	.3289	1.45	.4265
0.46	.1772	0.96	.3315	1.46	.4279
0.47	.1808	0.97	.3340	1.47	.4292
0.48	.1844	0.98	.3365	1.48	.4306
0.49	.1879	0.99	.3389	1.49	.4319

TABLE B *(continued)*

1.50	.4332	2.12	.4830	2.74	.4969
1.51	.4355	2.13	.4834	2.75	.4970
1.52	.4357	2.14	.4838	2.76	.4971
1.53	.4370	2.15	.4842	2.77	.4972
1.54	.4328	2.16	.4846	2.78	.4973
1.55	.4406	2.17	.4850	2.79	.4974
1.56	.4406	2.18	.4854	2.80	.4974
1.57	.4418	2.19	.4857	2.81	.4975
1.58	.4429	2.20	.4861	2.82	.4976
1.59	.4441	2.21	.4864	2.83	.4977
1.60	.4452	2.22	.4868	2.84	.4977
1.61	.4463	2.23	.4871	2.85	.4978
1.62	.4474	2.24	.4875	2.86	.4979
1.63	.4484	2.25	.4878	2.87	.4979
1.64	.4495	2.26	.4881	2.88	.4980
1.65	.4505	2.27	.4884	2.89	.4981
1.66	.4515	2.28	.4887	2.90	.4981
1.67	.4525	2.29	.4890	2.91	.4982
1.68	.4535	2.30	.4893	2.92	.4982
1.69	.4545	2.31	.4896	2.93	.4983
1.70	.4554	2.32	.4898	2.94	.4984
1.71	.4564	2.33	.4901	2.95	.4984
1.72	.4573	2.34	.4904	2.96	.4985
1.73	.4582	2.35	.4906	2.97	.4985
1.74	.4591	2.36	.4909	2.98	.4986
1.75	.4599	2.37	.4911	2.99	.4986
1.76	.4608	2.38	.4913	3.00	.4987
1.77	.4616	2.39	.4916	3.01	.4987
1.78	.4625	2.40	.4918	3.02	.4987
1.79	.4633	2.41	.4920	3.03	.4988
1.80	.4641	2.42	.4922	3.04	.4988
1.81	.4649	2.43	.4925	3.05	.4989
1.82	.4656	2.44	.4927	3.06	.4989
1.83	.4664	2.45	.4929	3.07	.4989
1.84	.4671	2.46	.4931	3.08	.4990
1.85	.4678	2.47	.4932	3.09	.4990
1.86	.4686	2.48	.4934	3.10	.4990
1.87	.4693	2.49	.4936	3.11	.4991
1.88	.4699	2.50	.4938	3.12	.4991
1.89	.4706	2.51	.4940	3.13	.4991
1.90	.4713	2.52	.4941	3.14	.4992
1.91	.4719	2.53	.4943	3.15	.4992
1.92	.4726	2.54	.4945	3.16	.4992
1.93	.4732	2.55	.4946	3.17	.4992
1.94	.4738	2.56	.4948	3.18	.4993
1.95	.4744	2.57	.4949	3.19	.4993
1.96	.4750	2.58	.4951	3.20	.4993
1.97	.4756	2.59	.4952	3.21	.4993
1.98	.4761	2.60	.4953	3.22	.4994
1.99	.4767	2.61	.4955	3.23	.4994
2.00	.4772	2.62	.4956	3.24	.4994
2.01	.4778	2.63	.4957	3.25	.4994
2.02	.4783	2.64	.4959	3.30	.4995
2.03	.4788	2.65	.4960	3.35	.4996
2.04	.4793	2.66	.4961	3.40	.4997
2.05	.4798	2.67	.4962	3.45	.4997
2.06	.4803	2.68	.4963	3.50	.4998
2.07	.4808	2.69	.4964	3.60	.4998
2.08	.4812	2.70	.4965	3.70	.4999
2.09	.4817	2.71	.4966	3.80	.4999
2.10	.4821	2.72	.4967	3.90	.49995
2.11	.4826	2.73	.4968	4.00	.49997

TABLE C
Critical Values of *t*

df	\alpha levels for a nondirectional (two-tailed) test					
	.2	.1	.05	.02	.01	.001
1	3.078	6.314	**12.706**	31.821	63.657	636.619
2	1.886	2.920	**4.303**	6.965	9.925	31.598
3	1.638	2.353	**3.182**	4.541	5.841	12.964
4	1.533	2.132	**2.776**	3.747	4.604	8.610
5	1.476	2.015	**2.571**	3.365	4.032	6.869
6	1.440	1.943	**2.447**	3.143	3.707	5.959
7	1.415	1.895	**2.365**	2.998	3.499	5.408
8	1.397	1.860	**2.306**	2.896	3.355	5.041
9	1.383	1.833	**2.262**	2.821	3.250	4.781
10	1.372	1.812	**2.228**	2.764	3.169	4.587
11	1.363	1.796	**2.201**	2.718	3.106	4.437
12	1.356	1.782	**2.179**	2.681	3.055	4.318
13	1.350	1.771	**2.160**	2.650	3.012	4.221
14	1.345	1.761	**2.145**	2.624	2.977	4.140
15	1.341	1.753	**2.131**	2.602	2.947	4.073
16	1.337	1.746	**2.120**	2.583	2.921	4.015
17	1.333	1.740	**2.110**	2.567	2.898	3.965
18	1.330	1.734	**2.101**	2.552	2.878	3.965
19	1.328	1.729	**2.093**	2.539	2.861	3.883
20	1.325	1.725	**2.086**	2.528	2.845	3.850
21	1.323	1.721	**2.080**	2.518	2.831	3.819
22	1.321	1.717	**2.074**	2.508	2.819	3.792
23	1.319	1.714	**2.069**	2.500	2.807	3.767
24	1.318	1.711	**2.064**	2.492	2.797	3.745
25	1.316	1.708	**2.060**	2.485	2.787	3.725
26	1.315	1.706	**2.056**	2.479	2.779	3.707
27	1.314	1.703	**2.052**	2.473	2.771	3.690
28	1.313	1.701	**2.048**	2.467	2.763	3.674
29	1.311	1.699	**2.045**	2.462	2.756	3.569
30	1.310	1.697	**2.042**	2.457	2.750	3.646
40	1.303	1.684	**2.021**	2.423	2.704	3.551
60	1.296	1.671	**2.000**	2.390	2.660	3.460
120	1.289	1.658	**1.980**	2.358	2.617	3.373
∞	1.282	1.645	**1.960**	2.326	2.576	3.291
	.1	.05	**.025**	.01	.005	.0005
	\alpha levels for a directional (one-tailed) test					

TABLE D
Percentage points of the *F*-distribution
Upper 10% points

df denominator	df numerator																		
	1	2	3	4	5	6	7	8	9	10	12	15	20	24	30	40	60	120	∞
1	39.86	49.50	53.59	55.83	57.24	58.20	58.91	59.44	59.86	60.19	60.71	61.22	61.74	62.00	62.26	62.53	62.79	63.06	63.33
2	8.53	9.00	9.16	9.24	9.29	9.33	9.35	9.37	9.38	9.39	9.41	9.42	9.44	9.45	9.46	9.47	9.47	9.48	9.49
3	5.54	5.46	5.39	5.34	5.31	5.28	5.27	5.25	5.24	5.23	5.22	5.20	5.18	5.18	5.17	5.16	5.15	5.14	5.13
4	4.54	4.32	4.19	4.11	4.05	4.01	3.98	3.95	3.94	3.92	3.90	3.87	3.84	3.83	3.82	3.80	3.79	3.78	3.76
5	4.06	3.78	3.62	3.52	3.45	3.40	3.37	3.34	3.32	3.30	3.27	3.24	3.21	3.19	3.17	3.16	3.14	3.12	3.10
6	3.78	3.46	3.29	3.18	3.11	3.05	3.01	2.98	2.96	2.94	2.90	2.87	2.84	2.82	2.80	2.78	2.76	2.74	2.72
7	3.59	3.26	3.07	2.96	2.88	2.83	2.78	2.75	2.72	2.70	2.67	2.63	2.59	2.58	2.56	2.54	2.51	2.49	2.47
8	3.46	3.11	2.92	2.81	2.73	2.67	2.62	2.59	2.56	2.54	2.50	2.46	2.42	2.40	2.38	2.36	2.34	2.32	2.29
9	3.36	3.01	2.81	2.69	2.61	2.55	2.51	2.47	2.44	2.42	2.38	2.34	2.30	2.28	2.25	2.23	2.21	2.18	2.16
10	3.29	2.92	2.73	2.61	2.52	2.46	2.41	2.38	2.35	2.32	2.28	2.24	2.20	2.18	2.16	2.13	2.11	2.08	2.06
11	3.23	2.86	2.66	2.54	2.45	2.39	2.34	2.30	2.27	2.25	2.21	2.17	2.12	2.10	2.08	2.05	2.03	2.00	1.97
12	3.18	2.81	2.61	2.48	2.39	2.33	2.28	2.24	2.21	2.19	2.15	2.10	2.06	2.04	2.01	1.99	1.96	1.93	1.90
13	3.14	2.76	2.56	2.43	2.35	2.28	2.23	2.20	2.16	2.14	2.10	2.05	2.01	1.98	1.96	1.93	1.90	1.88	1.85
14	3.10	2.73	2.52	2.39	2.31	2.24	2.19	2.15	2.12	2.10	2.05	2.01	1.96	1.94	1.91	1.89	1.86	1.83	1.80
15	3.07	2.70	2.49	2.36	2.27	2.21	2.16	2.12	2.09	2.06	2.02	1.97	1.92	1.90	1.87	1.85	1.82	1.79	1.76
16	3.05	2.67	2.46	2.33	2.24	2.18	2.13	2.09	2.06	2.03	1.99	1.94	1.89	1.87	1.84	1.81	1.78	1.75	1.72
17	3.03	2.64	2.44	2.31	2.22	2.15	2.10	2.06	2.03	2.00	1.96	1.91	1.86	1.84	1.81	1.78	1.75	1.72	1.69
18	3.01	2.62	2.42	2.29	2.20	2.13	2.08	2.04	2.00	1.98	1.93	1.89	1.84	1.81	1.78	1.75	1.72	1.69	1.66
19	2.99	2.61	2.40	2.27	2.18	2.11	2.06	2.02	1.98	1.96	1.91	1.86	1.81	1.79	1.76	1.73	1.70	1.67	1.63
20	2.97	2.59	2.38	2.25	2.16	2.09	2.04	2.00	1.96	1.94	1.89	1.84	1.79	1.77	1.74	1.71	1.68	1.64	1.61
21	2.96	2.57	2.36	2.23	2.14	2.08	2.02	1.98	1.95	1.92	1.87	1.83	1.78	1.75	1.72	1.69	1.66	1.62	1.59
22	2.95	2.56	2.35	2.22	2.13	2.06	2.01	1.97	1.93	1.90	1.86	1.81	1.76	1.73	1.70	1.67	1.64	1.60	1.57
23	2.94	2.55	2.34	2.21	2.11	2.05	1.99	1.95	1.92	1.89	1.84	1.80	1.74	1.72	1.69	1.66	1.62	1.59	1.55
24	2.93	2.54	2.33	2.19	2.10	2.04	1.98	1.94	1.91	1.88	1.83	1.78	1.73	1.70	1.67	1.64	1.61	1.57	1.53
25	2.92	2.53	2.32	2.18	2.09	2.02	1.97	1.93	1.89	1.87	1.82	1.77	1.72	1.69	1.66	1.63	1.59	1.56	1.52
26	2.91	2.52	2.31	2.17	2.08	2.01	1.96	1.92	1.88	1.86	1.81	1.76	1.71	1.68	1.65	1.61	1.58	1.54	1.50
27	2.90	2.51	2.30	2.17	2.07	2.00	1.95	1.91	1.87	1.85	1.80	1.75	1.70	1.67	1.64	1.60	1.57	1.53	1.49
28	2.89	2.50	2.29	2.16	2.06	2.00	1.94	1.90	1.87	1.84	1.79	1.74	1.69	1.66	1.63	1.59	1.56	1.52	1.48
29	2.89	2.50	2.28	2.15	2.06	1.99	1.93	1.89	1.86	1.83	1.78	1.73	1.68	1.65	1.62	1.58	1.55	1.51	1.47
30	2.88	2.49	2.28	2.14	2.05	1.98	1.93	1.88	1.85	1.82	1.77	1.72	1.67	1.64	1.61	1.57	1.54	1.50	1.46
40	2.84	2.44	2.23	2.09	2.00	1.93	1.87	1.83	1.79	1.76	1.71	1.66	1.61	1.57	1.54	1.51	1.47	1.42	1.38
60	2.79	2.39	2.18	2.04	1.95	1.87	1.82	1.77	1.74	1.71	1.66	1.60	1.54	1.51	1.48	1.44	1.40	1.35	1.29
120	2.75	2.35	2.13	1.99	1.90	1.82	1.77	1.72	1.68	1.65	1.60	1.55	1.48	1.45	1.41	1.37	1.32	1.26	1.19
∞	2.71	2.30	2.08	1.94	1.85	1.77	1.72	1.67	1.63	1.60	1.55	1.49	1.42	1.38	1.34	1.30	1.24	1.17	1.00

df numerator

df denominator	1	2	3	4	5	6	7	8	9	10	12	15	20	24	30	40	60	120	∞
1	161.4	199.5	215.7	224.6	230.2	234.0	236.8	238.9	240.5	241.9	243.9	245.9	248.0	249.1	250.1	251.1	252.2	253.3	254.3
2	18.51	19.00	19.16	19.25	19.30	19.33	19.35	19.37	19.38	19.40	19.41	19.43	19.45	19.46	19.47	19.48	19.49	19.50	19.50
3	10.13	9.55	9.28	9.12	9.01	8.94	8.89	8.85	8.81	8.79	8.74	8.70	8.66	8.64	8.62	8.59	8.57	8.55	8.53
4	7.71	6.94	6.59	6.39	6.26	6.16	6.09	6.04	6.00	5.96	5.91	5.86	5.80	5.77	5.75	5.72	5.69	5.66	5.63
5	6.61	5.79	5.41	5.19	5.05	4.95	4.88	4.82	4.77	4.74	4.68	4.62	4.56	4.53	4.50	4.46	4.43	4.40	4.36
6	5.99	5.14	4.76	4.53	4.39	4.28	4.21	4.15	4.10	4.06	4.00	3.94	3.87	3.84	3.81	3.77	3.74	3.70	3.67
7	5.59	4.74	4.35	4.12	3.97	3.87	3.79	3.73	3.68	3.64	3.57	3.51	3.44	3.41	3.38	3.34	3.30	3.27	3.23
8	5.32	4.46	4.07	3.84	3.69	3.58	3.50	3.44	3.39	3.35	3.28	3.22	3.15	3.12	3.08	3.04	3.01	2.97	2.93
9	5.12	4.26	3.86	3.63	3.48	3.37	3.29	3.23	3.18	3.14	3.07	3.01	2.94	2.90	2.86	2.83	2.79	2.75	2.71
10	4.96	4.10	3.71	3.48	3.33	3.22	3.14	3.07	3.02	2.98	2.91	2.85	2.77	2.74	2.70	2.66	2.62	2.58	2.54
11	4.84	3.98	3.59	3.36	3.20	3.09	3.01	2.95	2.90	2.85	2.79	2.72	2.65	2.61	2.57	2.53	2.49	2.45	2.40
12	4.75	3.89	3.49	3.26	3.11	3.00	2.91	2.85	2.80	2.75	2.69	2.62	2.54	2.51	2.47	2.43	2.38	2.34	2.30
13	4.67	3.81	3.41	3.18	3.03	2.92	2.83	2.77	2.71	2.67	2.60	2.53	2.46	2.42	2.38	2.34	2.30	2.25	2.21
14	4.60	3.74	3.34	3.11	2.96	2.85	2.76	2.70	2.65	2.60	2.53	2.46	2.39	2.35	2.31	2.27	2.22	2.18	2.13
15	4.54	3.68	3.29	3.06	2.90	2.79	2.71	2.64	2.59	2.54	2.48	2.40	2.33	2.29	2.25	2.20	2.16	2.11	2.07
16	4.49	3.63	3.24	3.01	2.85	2.74	2.66	2.59	2.54	2.49	2.42	2.35	2.28	2.24	2.19	2.15	2.11	2.06	2.01
17	4.45	3.59	3.20	2.96	2.81	2.70	2.61	2.55	2.49	2.45	2.38	2.31	2.23	2.19	2.15	2.10	2.06	2.01	1.96
18	4.41	3.55	3.16	2.93	2.77	2.66	2.58	2.51	2.46	2.41	2.34	2.27	2.19	2.15	2.11	2.06	2.02	1.97	1.92
19	4.38	3.52	3.13	2.90	2.74	2.63	2.54	2.48	2.42	2.38	2.31	2.23	2.16	2.11	2.07	2.03	1.98	1.93	1.88
20	4.35	3.49	3.10	2.87	2.71	2.60	2.51	2.45	2.39	2.35	2.28	2.20	2.12	2.08	2.04	1.99	1.95	1.90	1.84
21	4.32	3.47	3.07	2.84	2.68	2.57	2.49	2.42	2.37	2.32	2.25	2.18	2.10	2.05	2.01	1.96	1.92	1.87	1.81
22	4.30	3.44	3.05	2.82	2.66	2.55	2.46	2.40	2.34	2.30	2.23	2.15	2.07	2.03	1.98	1.94	1.89	1.84	1.78
23	4.28	3.42	3.03	2.80	2.64	2.53	2.44	2.37	2.32	2.27	2.20	2.13	2.05	2.01	1.96	1.91	1.86	1.81	1.76
24	4.26	3.40	3.01	2.78	2.62	2.51	2.42	2.36	2.30	2.25	2.18	2.11	2.03	1.98	1.94	1.89	1.84	1.79	1.73
25	4.24	3.39	2.99	2.76	2.60	2.49	2.40	2.34	2.28	2.24	2.16	2.09	2.01	1.96	1.92	1.87	1.82	1.77	1.71
26	4.23	3.37	2.98	2.74	2.59	2.47	2.39	2.32	2.27	2.22	2.15	2.07	1.99	1.95	1.92	1.85	1.80	1.75	1.69
27	4.21	3.35	2.96	2.73	2.57	2.46	2.37	2.31	2.25	2.20	2.13	2.06	1.97	1.93	1.88	1.84	1.79	1.73	1.67
28	4.20	3.34	2.95	2.71	2.56	2.45	2.36	2.29	2.24	2.19	2.12	2.04	1.96	1.91	1.87	1.82	1.77	1.71	1.65
29	4.18	3.33	2.93	2.70	2.55	2.43	2.35	2.28	2.22	2.18	2.10	2.03	1.94	1.90	1.85	1.81	1.75	1.70	1.64
30	4.17	3.32	2.92	2.69	2.53	2.42	2.33	2.27	2.21	2.16	2.09	2.01	1.93	1.89	1.84	1.79	1.74	1.68	1.62
40	4.08	3.23	2.84	2.61	2.45	2.34	2.25	2.18	2.12	2.08	2.00	1.92	1.84	1.79	1.74	1.69	1.64	1.58	1.51
60	4.00	3.15	2.76	2.53	2.37	2.25	2.17	2.10	2.04	1.99	1.92	1.84	1.75	1.70	1.65	1.59	1.53	1.47	1.39
120	3.92	3.07	2.68	2.45	2.29	2.17	2.09	2.02	1.96	1.91	1.83	1.75	1.66	1.61	1.55	1.50	1.43	1.35	1.25
∞	3.84	3.00	2.60	2.37	2.21	2.10	2.01	1.94	1.88	1.83	1.75	1.67	1.57	1.52	1.46	1.39	1.32	1.22	1.00

TABLE D *(continued)*
Upper 2.5% points

df numerator

df denominator	1	2	3	4	5	6	7	8	9	10	12	15	20	24	30	40	60	120	∞
1	647.8	799.5	864.2	899.6	921.8	937.1	948.2	956.7	963.3	968.6	976.7	984.9	993.1	997.2	1001	1006	1010	1014	1018
2	38.51	39.00	39.17	39.25	39.30	39.33	39.36	39.37	39.39	39.40	39.41	39.43	39.45	39.46	39.46	39.47	39.48	39.49	39.50
3	17.44	16.04	15.44	15.10	14.88	14.73	14.62	14.54	14.47	14.42	14.34	14.25	14.17	14.12	14.08	14.04	13.99	13.95	13.90
4	12.22	10.65	9.98	9.60	9.36	9.20	9.07	8.98	8.90	8.84	8.75	8.66	8.56	8.51	8.46	8.41	8.36	8.31	8.26
5	10.01	8.43	7.76	7.39	7.15	6.98	6.85	6.76	6.68	6.62	6.52	6.43	6.33	6.28	6.23	6.18	6.12	6.07	6.02
6	8.81	7.26	6.60	6.23	5.99	5.82	5.70	5.60	5.52	5.46	5.37	5.27	5.17	5.12	5.07	5.01	4.96	4.90	4.85
7	8.07	6.54	5.89	5.52	5.29	5.12	4.99	4.90	4.82	4.76	4.67	4.57	4.47	4.42	4.36	4.31	4.25	4.20	4.14
8	7.57	6.06	5.42	5.05	4.82	4.65	4.53	4.43	4.36	4.30	4.20	4.10	4.00	3.95	3.89	3.84	3.78	3.73	3.67
9	7.21	5.71	5.08	4.72	4.48	4.32	4.20	4.10	4.03	3.96	3.87	3.77	3.67	3.61	3.56	3.51	3.45	3.39	3.33
10	6.94	5.46	4.83	4.47	4.24	4.07	3.95	3.85	3.78	3.72	3.62	3.52	3.42	3.37	3.31	3.26	3.20	3.14	3.08
11	6.72	5.26	4.63	4.28	4.04	3.88	3.76	3.66	3.59	3.53	3.43	3.33	3.23	3.17	3.12	3.06	3.00	2.94	2.88
12	6.55	5.10	4.47	4.12	3.89	3.73	3.61	3.51	3.44	3.37	3.28	3.18	3.07	3.02	2.96	2.91	2.85	2.79	2.72
13	6.41	4.97	4.35	4.00	3.77	3.60	3.48	3.39	3.31	3.25	3.15	3.05	2.95	2.89	2.84	2.78	2.72	2.66	2.60
14	6.30	4.86	4.24	3.89	3.66	3.50	3.38	3.29	3.21	3.15	3.05	2.95	2.84	2.79	2.73	2.67	2.61	2.55	2.49
15	6.20	4.77	4.15	3.80	3.58	3.41	3.29	3.20	3.12	3.06	2.96	2.86	2.76	2.70	2.64	2.59	2.52	2.46	2.40
16	6.12	4.69	4.08	3.73	3.50	3.34	3.22	3.12	3.05	2.99	2.89	2.79	2.68	2.63	2.57	2.51	2.45	2.38	2.32
17	6.04	4.62	4.01	3.66	3.44	3.28	3.16	3.06	2.98	2.92	2.82	2.72	2.62	2.56	2.50	2.44	2.38	2.32	2.25
18	5.98	4.56	3.95	3.61	3.38	3.22	3.10	3.01	2.93	2.87	2.77	2.67	2.56	2.50	2.44	2.38	2.32	2.26	2.19
19	5.92	4.51	3.90	3.56	3.33	3.17	3.05	2.96	2.88	2.82	2.72	2.62	2.51	2.45	2.39	2.33	2.27	2.20	2.13
20	5.87	4.46	3.86	3.51	3.29	3.13	3.01	2.91	2.84	2.77	2.68	2.57	2.46	2.41	2.35	2.29	2.22	2.16	2.09
21	5.83	4.42	3.82	3.48	3.25	3.09	2.97	2.87	2.80	2.73	2.64	2.53	2.42	2.37	2.31	2.25	2.18	2.11	2.04
22	5.79	4.38	3.78	3.44	3.22	3.05	2.93	2.84	2.76	2.70	2.60	2.50	2.39	2.33	2.27	2.21	2.14	2.08	2.00
23	5.75	4.35	3.75	3.41	3.18	3.02	2.90	2.81	2.73	2.67	2.57	2.47	2.36	2.30	2.24	2.18	2.11	2.04	1.97
24	5.72	4.32	3.72	3.38	3.15	2.99	2.87	2.78	2.70	2.64	2.54	2.44	2.33	2.27	2.21	2.15	2.08	2.01	1.94
25	5.69	4.29	3.69	3.35	3.13	2.97	2.85	2.75	2.68	2.61	2.51	2.41	2.30	2.24	2.18	2.12	2.05	1.98	1.91
26	5.66	4.27	3.67	3.33	3.10	2.94	2.82	2.73	2.65	2.59	2.49	2.39	2.28	2.22	2.16	2.09	2.03	1.95	1.88
27	5.63	4.24	3.65	3.31	3.08	2.92	2.80	2.71	2.63	2.57	2.47	2.36	2.25	2.19	2.13	2.07	2.00	1.93	1.85
28	5.61	4.22	3.63	3.29	3.06	2.90	2.78	2.69	2.61	2.55	2.45	2.34	2.23	2.17	2.11	2.05	1.98	1.91	1.83
29	5.59	4.20	3.61	3.27	3.04	2.88	2.76	2.67	2.59	2.53	2.43	2.32	2.21	2.15	2.09	2.03	1.96	1.89	1.81
30	5.57	4.18	3.59	3.25	3.03	2.87	2.75	2.65	2.57	2.51	2.41	2.31	2.20	2.14	2.07	2.01	1.94	1.87	1.79
40	5.42	4.05	3.46	3.13	2.90	2.74	2.62	2.53	2.45	2.39	2.29	2.18	2.07	2.01	1.94	1.88	1.80	1.72	1.64
60	5.29	3.93	3.34	3.01	2.79	2.63	2.51	2.41	2.33	2.27	2.17	2.06	1.94	1.88	1.82	1.74	1.67	1.58	1.48
120	5.15	3.80	3.23	2.89	2.67	2.52	2.39	2.30	2.22	2.16	2.05	1.94	1.82	1.76	1.69	1.61	1.53	1.43	1.31
∞	5.02	3.69	3.12	2.79	2.57	2.41	2.29	2.19	2.11	2.05	1.94	1.83	1.71	1.64	1.57	1.48	1.39	1.27	1.00

TABLE D *(continued)*
Upper 1% points

df numerator

df denominator	1	2	3	4	5	6	7	8	9	10	12	15	20	24	30	40	60	120	∞
2	98.50	99.00	99.17	99.25	99.30	99.33	99.36	99.37	99.39	99.40	99.42	99.43	99.45	99.46	99.47	99.47	99.48	99.49	99.50
3	34.12	30.82	29.46	28.71	28.24	27.91	27.67	27.49	27.35	27.23	27.05	26.87	26.69	26.60	26.50	26.41	26.32	26.22	26.13
4	21.20	18.00	16.69	15.98	15.52	15.21	14.98	14.80	14.66	14.55	14.37	14.20	14.02	13.93	13.84	13.75	13.65	13.56	13.46
5	16.26	13.27	12.06	11.39	10.97	10.67	10.46	10.29	10.16	10.05	9.89	9.72	9.55	9.47	9.38	9.29	9.20	9.11	9.02
6	13.75	10.92	9.78	9.15	8.75	8.47	8.26	8.10	7.98	7.87	7.72	7.56	7.40	7.31	7.23	7.14	7.06	6.97	6.88
7	12.25	9.55	8.45	7.85	7.46	7.19	6.99	6.84	6.72	6.62	6.47	6.31	6.16	6.07	5.99	5.91	5.82	5.74	5.65
8	11.26	8.65	7.59	7.01	6.63	6.37	6.18	6.03	5.91	5.81	5.67	5.52	5.36	5.28	5.20	5.12	5.03	4.95	4.86
9	10.56	8.02	6.99	6.42	6.06	5.80	5.61	5.47	5.35	5.26	5.11	4.96	4.81	4.73	4.65	4.57	4.48	4.40	4.31
10	10.04	7.56	6.55	5.99	5.64	5.39	5.20	5.06	4.94	4.85	4.71	4.56	4.41	4.33	4.25	4.17	4.08	4.00	3.91
11	9.65	7.21	6.22	5.67	5.32	5.07	4.89	4.74	4.63	4.54	4.40	4.25	4.10	4.02	3.94	3.86	3.78	3.69	3.60
12	9.33	6.93	5.95	5.41	5.06	4.82	4.64	4.50	4.39	4.30	4.16	4.01	3.86	3.78	3.70	3.62	3.54	3.45	3.36
13	9.07	6.70	5.74	5.21	4.86	4.62	4.44	4.30	4.19	4.10	3.96	3.82	3.66	3.59	3.51	3.43	3.34	3.25	3.17
14	8.86	6.51	5.56	5.04	4.69	4.46	4.28	4.14	4.03	3.94	3.80	3.66	3.51	3.43	3.35	3.27	3.18	3.09	3.00
15	8.68	6.36	5.42	4.89	4.56	4.32	4.14	4.00	3.89	3.80	3.67	3.52	3.37	3.29	3.21	3.13	3.05	2.96	2.87
16	8.53	6.23	5.29	4.77	4.44	4.20	4.03	3.89	3.78	3.69	3.55	3.41	3.26	3.18	3.10	3.02	2.93	2.84	2.75
17	8.40	6.11	5.18	4.67	4.34	4.10	3.93	3.79	3.68	3.59	3.46	3.31	3.16	3.08	3.00	2.92	2.83	2.75	2.65
18	8.29	6.01	5.09	4.58	4.25	4.01	3.84	3.71	3.60	3.51	3.37	3.23	3.08	3.00	2.92	2.84	2.75	2.66	2.57
19	8.18	5.93	5.01	4.50	4.17	3.94	3.77	3.63	3.52	3.43	3.30	3.15	3.00	2.92	2.84	2.76	2.67	2.58	2.49
20	8.10	5.85	4.94	4.43	4.10	3.87	3.70	3.56	3.46	3.37	3.23	3.09	2.94	2.86	2.78	2.69	2.61	2.52	2.42
21	8.02	5.78	4.87	4.37	4.04	3.81	3.64	3.51	3.40	3.31	3.17	3.03	2.88	2.80	2.72	2.64	2.55	2.46	2.36
22	7.95	5.72	4.82	4.31	3.99	3.76	3.59	3.45	3.35	3.26	3.12	2.98	2.83	2.75	2.67	2.58	2.50	2.40	2.31
23	7.88	5.66	4.76	4.26	3.94	3.71	3.54	3.41	3.30	3.21	3.07	2.93	2.78	2.70	2.62	2.54	2.45	2.35	2.26
24	7.82	5.61	4.72	4.22	3.90	3.67	3.50	3.36	3.26	3.17	3.03	2.89	2.74	2.66	2.58	2.49	2.40	2.31	2.21
25	7.77	5.57	4.68	4.18	3.85	3.63	3.46	3.32	3.22	3.13	2.99	2.85	2.70	2.62	2.54	2.45	2.36	2.27	2.17
26	7.72	5.53	4.64	4.14	3.82	3.59	3.42	3.29	3.18	3.09	2.96	2.81	2.66	2.58	2.50	2.42	2.33	2.23	2.13
27	7.68	5.49	4.60	4.11	3.78	3.56	3.39	3.26	3.15	3.06	2.93	2.78	2.63	2.55	2.47	2.38	2.29	2.20	2.10
28	7.64	5.45	4.57	4.07	3.75	3.53	3.36	3.23	3.12	3.03	2.90	2.75	2.60	2.52	2.44	2.35	2.26	2.17	2.06
29	7.60	5.42	4.54	4.04	3.73	3.50	3.33	3.20	3.09	3.00	2.87	2.73	2.57	2.49	2.41	2.33	2.23	2.14	2.03
30	7.56	5.39	4.51	4.02	3.70	3.47	3.30	3.17	3.07	2.98	2.84	2.70	2.55	2.47	2.39	2.30	2.21	2.11	2.01
40	7.31	5.18	4.31	3.83	3.51	3.29	3.12	2.99	2.89	2.80	2.66	2.52	2.37	2.29	2.20	2.11	2.02	1.92	1.80
60	7.08	4.98	4.13	3.65	3.34	3.12	2.95	2.82	2.72	2.63	2.50	2.35	2.20	2.12	2.03	1.94	1.84	1.73	1.60
120	6.85	4.79	3.95	3.48	3.17	2.96	2.79	2.66	2.56	2.47	2.34	2.19	2.03	1.95	1.86	1.76	1.66	1.53	1.38
∞	6.63	4.61	3.78	3.32	3.02	2.80	2.64	2.51	2.41	2.32	2.18	2.04	1.88	1.79	1.70	1.59	1.47	1.32	1.00

df numerator

df denominator	1	2	3	4	5	6	7	8	9	10	12	15	20	24	30	40	60	120	∞
2	198.5	199.0	199.2	199.2	199.3	199.3	199.4	199.4	199.4	199.4	199.4	199.4	199.4	199.5	199.5	199.5	199.5	199.5	199.5
3	55.55	49.80	47.47	46.19	45.39	44.84	44.43	44.13	43.88	43.69	43.39	43.08	42.78	42.62	42.47	42.31	42.15	41.99	41.83
4	31.33	26.28	24.26	23.15	22.46	21.97	21.62	21.35	21.14	20.97	20.70	20.44	20.17	20.03	19.89	19.75	19.61	19.47	19.32
5	22.78	18.31	16.53	15.56	14.94	14.51	14.20	13.96	13.77	13.62	13.38	13.15	12.90	12.78	12.66	12.53	12.40	12.27	12.14
6	18.63	14.54	12.92	12.03	11.46	11.07	10.79	10.57	10.39	10.25	10.03	9.81	9.59	9.47	9.36	9.24	9.12	9.00	8.88
7	16.24	12.40	10.88	10.05	9.52	9.16	8.89	8.68	8.51	8.38	8.18	7.97	7.75	7.65	7.53	7.42	7.31	7.19	7.08
8	14.69	11.04	9.60	8.81	8.30	7.95	7.69	7.50	7.34	7.21	7.01	6.81	6.61	6.50	6.40	6.29	6.18	6.06	5.95
9	13.61	10.11	8.72	7.96	7.47	7.13	6.88	6.69	6.54	6.42	6.23	6.03	5.83	5.73	5.62	5.52	5.41	5.30	5.19
10	12.83	9.43	8.08	7.34	6.87	6.54	6.30	6.12	5.97	5.85	5.66	5.47	5.27	5.17	5.07	4.97	4.86	4.75	4.64
11	12.23	8.91	7.60	6.88	6.42	6.10	5.86	5.68	5.54	5.42	5.24	5.05	4.86	4.76	4.65	4.55	4.44	4.34	4.23
12	11.75	8.51	7.23	6.52	6.07	5.76	5.52	5.35	5.20	5.09	4.91	4.72	4.53	4.43	4.33	4.23	4.12	4.01	3.90
13	11.37	8.19	6.93	6.23	5.79	5.48	5.25	5.08	4.94	4.82	4.64	4.46	4.27	4.17	4.07	3.97	3.87	3.76	3.65
14	11.06	7.92	6.68	6.00	5.56	5.26	5.03	4.86	4.72	4.60	4.43	4.25	4.06	3.96	3.86	3.76	3.66	3.55	3.44
15	10.80	7.70	6.48	5.80	5.37	5.07	4.85	4.67	4.54	4.42	4.25	4.07	3.88	3.79	3.69	3.58	3.48	3.37	3.26
16	10.58	7.51	6.30	5.64	5.21	4.91	4.69	4.52	4.38	4.27	4.10	3.92	3.73	3.64	3.54	3.44	3.33	3.22	3.11
17	10.38	7.35	6.16	5.50	5.07	4.78	4.56	4.39	4.25	4.14	3.97	3.79	3.61	3.51	3.41	3.31	3.21	3.10	2.98
18	10.22	7.21	6.03	5.37	4.96	4.66	4.44	4.28	4.14	4.03	3.86	3.68	3.50	3.40	3.30	3.20	3.10	2.99	2.87
19	10.07	7.09	5.92	5.27	4.85	4.56	4.34	4.18	4.04	3.93	3.76	3.59	3.40	3.31	3.21	3.11	3.00	2.89	2.78
20	9.94	6.99	5.82	5.17	4.76	4.47	4.26	4.09	3.96	3.85	3.68	3.50	3.32	3.22	3.12	3.02	2.92	2.81	2.69
21	9.83	6.89	5.73	5.09	4.68	4.39	4.18	4.01	3.88	3.77	3.60	3.43	3.24	3.15	3.05	2.95	2.84	2.73	2.61
22	9.73	6.81	5.65	5.02	4.61	4.32	4.11	3.94	3.81	3.70	3.54	3.36	3.18	3.08	2.98	2.88	2.77	2.66	2.55
23	9.63	6.73	5.58	4.95	4.54	4.26	4.05	3.88	3.75	3.64	3.47	3.30	3.12	3.02	2.92	2.82	2.71	2.60	2.48
24	9.55	6.66	5.52	4.89	4.49	4.20	3.99	3.83	3.69	3.59	3.42	3.25	3.06	2.97	2.87	2.77	2.66	2.55	2.43
25	9.48	6.60	5.46	4.84	4.43	4.15	3.94	3.78	3.64	3.54	3.37	3.20	3.01	2.92	2.82	2.72	2.61	2.50	2.38
26	9.41	6.54	5.41	4.79	4.38	4.10	3.89	3.73	3.60	3.49	3.33	3.15	2.97	2.87	2.77	2.67	2.56	2.45	2.33
27	9.34	6.49	5.36	4.74	4.34	4.06	3.85	3.69	3.56	3.45	3.28	3.11	2.93	2.83	2.73	2.63	2.52	2.41	2.29
28	9.28	6.44	5.32	4.70	4.30	4.02	3.81	3.65	3.52	3.41	3.25	3.07	2.89	2.79	2.69	2.59	2.48	2.37	2.25
29	9.23	6.40	5.28	4.66	4.26	3.98	3.77	3.61	3.48	3.38	3.21	3.04	2.86	2.76	2.66	2.56	2.45	2.33	2.21
30	9.18	6.35	5.24	4.62	4.23	3.95	3.74	3.58	3.45	3.34	3.18	3.01	2.82	2.73	2.63	2.52	2.42	2.30	2.18
40	8.83	6.07	4.98	4.37	3.99	3.71	3.51	3.35	3.22	3.12	2.95	2.78	2.60	2.50	2.40	2.30	2.18	2.06	1.93
60	8.49	5.79	4.73	4.14	3.76	3.49	3.29	3.13	3.01	2.90	2.74	2.57	2.39	2.29	2.19	2.08	1.96	1.83	1.69
120	8.18	5.54	4.50	3.92	3.55	3.28	3.09	2.93	2.81	2.71	2.54	2.37	2.19	2.09	1.98	1.87	1.75	1.61	1.43
∞	7.88	5.30	4.28	3.72	3.35	3.09	2.90	2.74	2.62	2.52	2.36	2.19	2.00	1.90	1.79	1.67	1.53	1.36	1.00

df numerator

df denominator	1	2	3	4	5	6	7	8	9	10	12	15	20	24	30	40	60	120	∞
2	998.5	999.0	999.2	999.2	999.3	999.3	999.4	999.4	999.4	999.4	999.4	999.4	999.4	999.5	999.5	999.5	999.5	999.5	999.5
3	167.0	148.5	141.1	137.1	134.6	132.8	131.6	130.6	129.9	129.2	128.3	127.4	126.4	125.9	125.4	125.0	124.5	124.0	123.5
4	74.14	61.25	56.18	53.44	51.71	50.53	49.66	49.00	48.47	48.05	47.41	46.76	46.10	45.77	45.43	45.09	44.75	44.40	44.05
5	47.18	37.12	33.20	31.09	29.75	28.84	28.16	27.64	27.24	26.92	26.42	25.91	25.39	25.14	24.87	24.60	24.33	24.06	23.79
6	35.51	27.00	23.70	21.92	20.81	20.03	19.46	19.03	18.69	18.41	17.99	17.56	17.12	16.89	16.67	16.44	16.21	15.99	15.75
7	29.25	21.69	18.77	17.19	16.21	15.52	15.02	14.63	14.33	14.08	13.71	13.32	12.93	12.73	12.53	12.33	12.12	11.91	11.70
8	25.42	18.49	15.83	14.39	13.49	12.86	12.40	12.04	11.77	11.54	11.19	10.84	10.48	10.30	10.11	9.92	9.73	9.53	9.33
9	22.86	16.39	13.90	12.56	11.71	11.13	10.70	10.37	10.11	9.89	9.57	9.24	8.90	8.72	8.55	8.37	8.19	8.00	7.81
10	21.04	14.91	12.55	11.28	10.48	9.92	9.52	9.20	8.96	8.75	8.45	8.13	7.80	7.64	7.47	7.30	7.12	6.94	6.76
11	19.69	13.81	11.56	10.35	9.58	9.05	8.66	8.35	8.12	7.92	7.63	7.32	7.01	6.85	6.68	6.52	6.35	6.17	6.00
12	18.64	12.97	10.80	9.63	8.89	8.38	8.00	7.71	7.48	7.29	7.00	6.71	6.40	6.25	6.09	5.93	5.76	5.59	5.42
13	17.81	12.31	10.21	9.07	8.35	7.86	7.49	7.21	6.98	6.80	6.52	6.23	5.93	5.78	5.63	5.47	5.30	5.14	4.97
14	17.14	11.78	9.73	8.62	7.92	7.43	7.08	6.80	6.58	6.40	6.13	5.85	5.56	5.41	5.25	5.10	4.94	4.77	4.60
15	16.59	11.34	9.34	8.25	7.57	7.09	6.74	6.47	6.26	6.08	5.81	5.54	5.25	5.10	4.95	4.80	4.64	4.47	4.31
16	16.12	10.97	9.00	7.94	7.27	6.81	6.46	6.19	5.98	5.81	5.55	5.27	4.99	4.85	4.70	4.54	4.39	4.23	4.06
17	15.72	10.66	8.73	7.68	7.02	6.56	6.22	5.96	5.75	5.58	5.32	5.05	4.78	4.63	4.48	4.33	4.18	4.02	3.85
18	15.38	10.39	8.49	7.46	6.81	6.35	6.02	5.76	5.56	5.39	5.13	4.87	4.59	4.45	4.30	4.15	4.00	3.84	3.67
19	15.08	10.16	8.28	7.26	6.62	6.18	5.85	5.59	5.39	5.22	4.97	4.70	4.43	4.29	4.14	3.99	3.84	3.68	3.51
20	14.82	9.95	8.10	7.10	6.46	6.02	5.69	5.44	5.24	5.08	4.82	4.56	4.29	4.15	4.00	3.86	3.70	3.54	3.38
21	14.59	9.77	7.94	6.95	6.32	5.88	5.56	5.31	5.11	4.95	4.70	4.44	4.17	4.03	3.88	3.74	3.58	3.42	3.26
22	14.38	9.61	7.80	6.81	6.19	5.76	5.44	5.19	4.99	4.83	4.58	4.33	4.06	3.92	3.78	3.63	3.48	3.32	3.15
23	14.19	9.47	7.67	6.69	6.08	5.65	5.33	5.09	4.89	4.73	4.48	4.23	3.96	3.82	3.68	3.53	3.38	3.22	3.05
24	14.03	9.34	7.55	6.59	5.98	5.55	5.23	4.99	4.80	4.64	4.39	4.14	3.87	3.74	3.59	3.45	3.29	3.14	2.97
25	13.88	9.22	7.45	6.49	5.88	5.46	5.15	4.91	4.71	4.56	4.31	4.06	3.79	3.66	3.52	3.37	3.22	3.06	2.89
26	13.74	9.12	7.36	6.41	5.80	5.38	5.07	4.83	4.64	4.48	4.24	3.99	3.72	3.59	3.44	3.30	3.15	2.99	2.82
27	13.61	9.02	7.27	6.33	5.73	5.31	5.00	4.76	4.57	4.41	4.17	3.92	3.66	3.52	3.38	3.23	3.08	2.92	2.75
28	13.50	8.93	7.19	6.25	5.66	5.24	4.93	4.69	4.50	4.35	4.11	3.86	3.60	3.46	3.32	3.18	3.02	2.86	2.69
29	13.39	8.85	7.12	6.19	5.59	5.18	4.87	4.64	4.45	4.29	4.05	3.80	3.54	3.41	3.27	3.12	2.97	2.81	2.64
30	13.29	8.77	7.05	6.12	5.53	5.12	4.82	4.58	4.39	4.24	4.00	3.75	3.49	3.36	3.22	3.07	2.92	2.76	2.59
40	12.61	8.25	6.60	5.70	5.13	4.73	4.44	4.21	4.02	3.87	3.64	3.40	3.15	3.01	2.87	2.73	2.57	2.41	2.23
60	11.97	7.76	6.17	5.31	4.76	4.37	4.09	3.87	3.69	3.54	3.31	3.08	2.83	2.69	2.55	2.41	2.25	2.08	1.89
120	11.38	7.32	5.79	4.95	4.42	4.04	3.77	3.55	3.38	3.24	3.02	2.78	2.53	2.40	2.26	2.11	1.95	1.76	1.54
∞	10.83	6.91	5.42	4.62	4.10	3.74	3.47	3.27	3.10	2.96	2.74	2.51	2.27	2.13	1.99	1.84	1.66	1.45	1.00

df denominator

TABLE E
Critical Values of Chi-Square

α values

df	0.250	0.100	0.050	0.025	0.010	0.005	0.001
1	1.32330	2.70554	3.84146	5.02389	6.63490	7.87944	10.828
2	2.77259	4.60517	5.99146	7.37776	9.21034	10.5966	13.816
3	4.10834	6.25139	7.81473	9.34840	11.3449	12.8382	16.266
4	5.38527	7.77944	9.48773	11.1433	13.2767	14.8603	18.467
5	6.62568	9.23636	11.0705	12.8325	15.0863	16.7496	20.515
6	7.84080	10.6446	12.5916	14.4494	16.8119	18.5476	22.458
7	9.03715	12.0170	14.0671	16.0128	18.4753	20.2777	24.322
8	10.2189	13.3616	15.5073	17.5345	20.0902	21.9550	26.125
9	11.3888	14.6837	16.9190	19.0228	21.6660	23.5894	27.877
10	12.5489	15.9872	18.3070	20.4832	23.2093	25.1882	29.588
11	13.7007	17.2750	19.6751	21.9200	24.7250	26.7568	31.264
12	14.8454	18.5493	21.0261	23.3367	26.2170	28.2995	32.909
13	15.9839	19.8119	22.3620	24.7356	27.6882	29.8195	34.528
14	17.1169	21.0641	23.6848	26.1189	29.1412	31.3194	36.123
15	18.2451	22.3071	24.9958	27.4884	30.5779	32.8013	37.697
16	19.3689	23.5418	26.2962	28.8454	31.9999	34.2672	39.252
17	20.4887	24.7690	27.5871	30.1910	33.4087	35.7185	40.790
18	21.6049	25.9894	28.8693	31.5264	34.8053	37.1565	42.312
19	22.7178	27.2036	30.1435	32.8523	36.1909	38.5823	43.820
20	23.8277	28.4120	31.4104	34.1696	37.5662	39.9968	45.315
21	24.9348	29.6151	32.6706	35.4789	38.9322	41.4011	46.797
22	26.0393	30.8133	33.9244	36.7807	40.2894	42.7957	48.268
23	27.1413	32.0069	35.1725	38.0756	41.6384	44.1813	49.728
24	28.2412	33.1962	36.4150	39.3641	42.9798	45.5585	51.179
25	29.3389	34.3816	37.6525	40.6465	44.3141	46.9279	52.618
26	30.4346	35.5632	38.8851	41.9232	45.6417	48.2899	54.052
27	31.5284	36.7412	40.1133	43.1945	46.9629	49.6449	55.476
28	32.6205	37.9159	41.3371	44.4608	48.2782	50.9934	56.892
29	33.7109	39.0875	42.5570	45.7223	49.5879	52.3356	58.301
30	34.7997	40.2560	43.7730	46.9792	50.8922	53.6720	59.703
40	45.6160	51.8051	55.7585	59.3417	63.6907	66.7660	73.402
50	56.3336	63.1671	67.5048	71.4202	76.1539	79.4900	86.661
60	66.9815	74.3970	79.0819	83.2977	88.3794	91.9517	99.607
70	77.5767	85.5270	90.5312	95.0232	100.425	104.215	112.317
80	88.1303	96.5782	101.879	106.629	112.329	116.321	124.839
90	98.6499	107.565	113.145	118.136	124.116	128.299	137.208
100	109.141	118.498	124.342	129.561	135.807	140.169	149.449

Appendix C

Answers to Selected Exercises

CHAPTER 1

EXPERIMENT 1

(a) I.V. is type of music.
(b) D.V. is driving performance of teenagers.
(c) Control condition is no music.
(d) Experimental conditions are classical and rock music.
(e) I.V. operationalized: No music condition, classical music by Bach and rock music from radio station 44 rock played at a 60-decibel level 5 feet from subject. D.V. operationalized: Recording the number of errors a driver makes on a driving simulator under one of the above I.V. conditions in a 10-minute time frame.
(f) I.V. scale is nominal; D.V. scale is ratio.
(b) Not confounded.
(h) Prototype experiment.
(i) Hypothesis: listening to rock music impedes the driving performance of teenagers.

EXPERIMENT 4

(a) Time-out procedure (i.e., interventions/no interventions).
(b) David's temper tantrums.
(c) Baseline number of tantrums.
(d) Time-out procedure.
(e) I.V. operationalized: No treatment during pretreatment week, time-out procedure during treatment week. D.V. operationalized: The number of tantrums David threw in a given time period.
(f) I.V. scale is nominal; D.V. scale is ratio.
(g) Confounded with time factors.
(h) Baseline or pre-post design.

(i) Hypothesis: a time-out procedure will stop 4-year-old David from throwing tantrums.

2a. Potentially testable, but probably not currently testable.

CHAPTER 3

1. mean = 10.8
   median = 9
   mode = 10
3. 7
7(a) median: few wealthy people would increase mean
 (b) mean or median: few very high and very low scores
9(a)

X	$X - \bar{X}$	$(X - \bar{X})^2$
3	-2	4
1	-4	16
8	3	9
9	4	16
5	0	0
4	-1	1
$\bar{X} = 5$		$46 = \Sigma(X - \bar{X})^2$
$N = 6$		

$$\sigma = \sqrt{\frac{46}{6}} = 2.77$$

11(a)

X	$X - \bar{X}$	$(X - \bar{X})^2$	X	$X - \bar{X}$	$(X - \bar{X})^2$
21	1	1	105	5.25	27.56
19	-1	1	99	$-.75$	0.56
30	10	100	94	-5.75	33.06
10	-10	100	101	1.25	1.56
15	-5	25	100	0.25	0.06
25	5	25	108	8.25	68.06
20	0	0	103	3.25	10.56
			88	-11.75	138.06

$\bar{X} = 20 \qquad 252 = \Sigma (X - \bar{X})^2 \qquad \bar{X} = 99.75 \qquad 279.48 = \Sigma (X - \bar{X})^2$

$$S = \sqrt{\frac{\Sigma (X - \bar{X})^2}{N - 1}} \qquad\qquad S = \sqrt{\frac{\Sigma (X - \bar{X})^2}{N - 1}}$$

$$S = \sqrt{\frac{252}{6}} \qquad\qquad S = \sqrt{\frac{279.48}{7}}$$

$$S = 6.48 \qquad\qquad S = 6.32$$

(c)

X	$X - \bar{X}$	$(X - \bar{X})^2$
35	10.25	105.06
17	−7.75	60.06
21	−3.75	14.06
28	3.25	10.56
31	6.25	39.06
19	−5.75	33.06
14	−10.75	115.56
33	8.25	68.06
$\bar{X} = 24.75$		$445.48 = \Sigma (X - \bar{X})^2$

$$S = \sqrt{\frac{\Sigma(X - \bar{X})^2}{N - 1}}$$

$$S = \sqrt{\frac{445.48}{7}}$$

$$S = 7.98$$

12(a)

X	X^2
21	441
19	361
30	900
10	100
15	225
25	625
20	400
140	3052

$$S = \sqrt{\frac{7(3052) - 19600}{7(6)}}$$

$$S = 6.48$$

14

X	$X - \bar{X}$	$(X - \bar{X})^2$
5	−2.14	4.58
7	−.14	.02
9	1.86	3.46
4	−3.14	9.86
11	3.86	14.90
1	−6.14	37.70
13	5.86	34.34
$\bar{X} = 7.14$		104.86

$$\sigma = \sqrt{\frac{\Sigma (X - \bar{X})^2}{N}} =$$

$$\sqrt{\frac{104.86}{7}} =$$

3.87

18	X	X^2
	28	784
	32	1024
	30	900
	34	1156
	29	841
	30	900
	30	900
	29	841
	33	1089
	31	961

$\bar{X} = 30.6 \qquad \Sigma X = 306 \qquad \Sigma X^2 = 9396.$

$$S = \sqrt{\frac{10(9396) - (306)^2}{(10)(9)}}$$

$$= 1.90$$

CHAPTER 4

1(a) $\dfrac{2}{6}$

1(c) $\dfrac{4}{6}$

1(e) $\dfrac{6}{6}$

3. $p(7H \text{ or } 8H \text{ or } 9H) = \dfrac{36}{512} + \dfrac{9}{512} + \dfrac{1}{512} = \dfrac{46}{512}$

5. $1 - p(6H \text{ or } 7H) = 1 - \left[\dfrac{7}{128} + \dfrac{1}{128} \right] = \dfrac{120}{128}$

6(c) $\left[\dfrac{1}{2} \right] \left[\dfrac{1}{2} \right] = \dfrac{1}{4}$

(e) $\dfrac{C_6^7 + C_7^7}{2^7} = \dfrac{7}{128} + \dfrac{1}{128} = \dfrac{8}{128}$

(i) $\dfrac{13}{52} + \dfrac{4}{52} - \dfrac{1}{52} = \dfrac{16}{52}$

7(a) $\dfrac{C_4^8 + C_4^8}{2^8} = \dfrac{70}{256} + \dfrac{70}{256} = \dfrac{140}{256}$

(c) $\dfrac{C_8^8 + C_0^8}{2^8} = \dfrac{1}{256} + \dfrac{1}{256} = \dfrac{2}{256}$

7b(a) $\dfrac{126}{512} + \dfrac{126}{512} = \dfrac{252}{512}$

(c) $\dfrac{C_8^9 + C_9^9 + C_0^9}{2^9} = \dfrac{11}{512}$

9.

directional test	nondirectional test
$\dfrac{C_8^8}{2^8} = \dfrac{1}{256} < .05$	$\dfrac{2}{256} < .05$
$\dfrac{C_8^8 + C_7^8}{2^8} = \dfrac{9}{256} < .05$	$\dfrac{18}{256} > .05$
$\dfrac{C_8^8 + C_7^8 + C_6^8}{2^8} = \dfrac{37}{256} > .05$	

Cutoff is between 6 and 7 for directional test and between 7 and 8 for nondirectional test.

CHAPTER 5

1(a)

$$X: \sigma = \sqrt{\dfrac{80}{5}} \qquad Z: \sigma = \sqrt{\dfrac{5}{5}}$$

$$\sigma = 4 \qquad\qquad \sigma = 1$$

3.   Yes—dependent on σ value

7.

Pam	Lois
$T = 57.3$	$T = 40$
$SAT = 573$	$SAT = 400$
$Z = .73$	$Z = -1$

8.

	X	μ	σ	Z score	T score	SAT score
	31	35	10	−.4	46	460

10(a)   .50 + .3413 = .8413 or 84.13%
  (b)   .50 − .4772 = .0228 or 2.28%
  (d)   .4772 − .3413 = .1359 or 13.59%

11(a)   .50 + .4082 = .9082
  (b)   .50 − .4082 = .0918
  (d)   .50 + .0927 = .5927

13.

Target area is between 1 and −1σ

.3413 + .3413 = .6826 or 68.26%

14(a) $\bar{X} = 13.5$
  $S = 3.42$
  $N = 4$
  $S_{\bar{X}} = \dfrac{3.42}{\sqrt{4}}$
  $= 1.71$

16(a)   11.26
17(a) $Z = 2.5 = .4938;\ .50 − .4938 = .0062$ or .62%
18.

$$S_{\bar{X}} = \frac{S}{\sqrt{N}} \qquad\qquad Z = \frac{2.46 − 0}{.49}$$

$$S_{\bar{X}} = \frac{3.1}{\sqrt{40}} \qquad\qquad Z = 5.02$$

$S_{\bar{X}} = .49$ 5.02 > 1.65 ∴ null hypothesis rejected

Note: 1.65 is critical Z value at $\alpha = .05$, directional test.

CHAPTER 6

1. $$t = \frac{-3}{\sqrt{\dfrac{5(55) - 225}{25(4)}}}$$

$$t = \frac{-3}{\sqrt{.5}}$$

$|t| = 4.24$ (ignoring the sign)

Critical $t = 2.776$ for nondirectional test; df = 4; $\alpha = .05$; 4.24 > 2.776 ∴ reject null hypothesis.

2. $$t = \frac{-3}{\sqrt{\dfrac{10}{5} + \dfrac{6}{5}}}$$

$|t| = 1.68$ (ignoring the sign)

Critical $t = 2.306$ for nondirectional test; df $= 8$; $\alpha = .05$; $1.68 < 2.306$ ∴ retain null hypothesis.

7.

(a)

$$t = \frac{15 - 19}{\sqrt{\dfrac{10}{6} + \dfrac{13}{6}}}$$

$|t| = 2.04$

Critical t (10 df), $\alpha = .05$, for nondirectional test is 2.228

$2.04 < 2.228$ ∴ retain H_0;

(b)

$$t = \frac{15 - 19}{\sqrt{\dfrac{10}{36} + \dfrac{13}{36}}}$$

$|t| = 5.00$

Critical t (70 df), $\alpha = .05$, for nondirectional test is 2.00

$5.00 > 2.00$ ∴ reject H_0.

10.   Independent group t test

$$t = \frac{73.4 - 84.9}{\sqrt{\dfrac{220.27}{10} + \dfrac{86.99}{10}}}$$

$|t| = 2.08$

Critical t (18 df), $\alpha = .05$, is 2.101 for nondirectional test. $2.07 < 2.101$ ∴ H_0 retained.

12.   Independent group t test, nondirectional, $\alpha = .05$, df $= 18$
Critical $t = 2.101$

$$t = \frac{160.7 - 150}{\sqrt{\dfrac{1083.57}{10} + \dfrac{972.67}{10}}}$$

$|t| = .746$

$.746 < 2.101$ ∴ retain H_0; no evidence that feedback influences amount of strength exerted in a strength task.

14.   Independent group t test, $\alpha = .05$, nondirectional test, df $= 28$
Critical $t = 2.048$

$$t = \frac{6.07 - 3.93}{\sqrt{\dfrac{25.78}{15} + \dfrac{15.07}{15}}}$$

$|t| = 1.30$

1.30 < 2.048 ∴ retain H_0; no evidence for difference between biofeedback group and control group in ability to lower their blood pressure.

CHAPTER 7

2.

$$MS_B = \frac{34.66}{2} = 17.33$$

$$MS_W = \frac{50}{9} = 5.56$$

$$F = \frac{17.33}{5.56} = 3.12$$

critical $F_{2,9}$ ($\alpha = .05$) = 4.26
3.12 < 4.26 ∴ retain H_0

6.

Source of Variation	SS	df	MS	F	p
Between	3.1	3	1.03	34.33	< .05
Within	.5	16	.03		
Total	3.6	19			

(critical F = 3.24)

$$LSD = 2.120 \sqrt{(.03)\left[\frac{2}{5}\right]} = .23$$

A:B $|2.1 - 1.7| > .23$, significant
A:C $|2.1 - 2.2| < .23$, nonsignificant
A:D $|2.1 - 2.8| > .23$, significant
B:C $|1.7 - 2.2| > .23$, significant
B:D $|1.7 - 2.8| > .23$, significant
C:D $|2.2 - 2.8| > .23$, significant

9.

Source of Variation	SS	df	MS	F	p
Between	68.25	2	34.13	15.44	< .05
Within	46.37	21	2.21		
Total	114.62	23			

Critical F = 3.47
Protected t tests

$$LSD = 2.08 \sqrt{2.21 \left[\frac{2}{8} \right]} = 1.54$$

A:B $|7.5 - 3.75| > 1.54$, significant
A:C $|7.5 - 7.13| < 1.54$, nonsignificant
B:C $|3.75 - 7.13| > 1.54$, significant

CHAPTER 8

1(a)

	A	B	C
M	10	9	8
F	14	7	12

Source of Variation	SS	df	MS	F	p
Rows	18	1	18	5.68	$< .05$
Columns	48	2	24	7.58	$< .05$
Interaction	36	2	18	5.68	$< .05$
Error	38	12	3.16		
Total	140	17			

Critical F's
F_{row} $(1,12) = 4.75$
F_{column} $(2,12) = 3.88$
$F_{interaction}$ $(2,12) = 3.88$

3(a)

	P	D
H	95	25
N	25	55

Source of Variation	SS	df	MS	F	p
Child type	2,000	1	2,000	23.26	$< .05$
Drug dosage	2,000	1	2,000	23.26	$< .05$
Interaction	12,500	1	12,500	145.35	$< .05$
Error	1,376	16	86		
Total	17,876	19			

Critical $F(1,16) = 4.49$ ($\alpha = .05$)

CHAPTER 9

2.

Source of Variation	SS	df	MS	F	p
SS_T	102.5	9			
SS_B	62.5	1	62.5	8.93	< .05
SS_W	40.0	8	5.0		
SS_S	12.0	4	3.0		
SS_{WS}	28.0	4	7.0		

yes, $t^2 = F$; $([2.99]^2 = 8.93)$

6(a)

		Retrieval Measure	
		Immed. Recall	2-week Recall
Word	E	14.00	13.25
Type	N	15.25	10.00

(b) possibly

(c)

Source of Variation	SS	df	MS	F	p
Total	159.75	15			
Between subjects	93.75	7			
rows (type words)	4.0	1	4	.267	> .05
error (S/G)	89.75	6	14.96		
Within subjects	66.00	8			
columns (type recall)	36.0	1	36.00	22.09	< .05
interaction	20.25	1	20.25	12.423	< .05
error (WS)	9.76	6	1.63		

(d) yes, the test for interaction

(e)

$$LSD_{within} = 2.447 \sqrt{1.63 \left[\frac{2}{4}\right]} \qquad LSD_{btwn} = 2.446 \sqrt{14.96 \left[\frac{2}{4}\right]}$$

$$= 2.21 \qquad\qquad\qquad = 6.69$$

EI:E2W $|69.00 - 13.25| > 2.21 \therefore$ reject H_0
EI:NI $|19 - 16.25| < 6.69 \therefore$ retain H_0
NI:N2W $|16.25 - 10.00| > 2.21 \therefore$ reject H_0
NI:E2W $|16.25 - 13.25| < 6.69 \therefore$ retain H_0

(f) Emotional words were better retained over a two-week period than were neutral words.

CHAPTER 10

Set 1

1b. .79

1c. .87

1d. -5.52

1e. $X = 10\ Y' = 3.18$
    $X = 15,\ Y' = 7.53$

1g. 1.51

3.   Set 1
    $X = 5:\ \ Y' = 1.17$
    $X = 10:\ Y' = 3.18$

4.   Set 1
    Critical t (5 df) $= 2.57$, H_0: population r is zero
    Computed $t = 4.65$
    $4.65 > 2.57\ \therefore$ reject H_0

7(a)   Computed $t = \dfrac{.45\sqrt{8}}{\sqrt{1 - .20}} = 1.43$

      Critical $t = 2.306$
      $1.43 < 2.306\ \therefore$ retain H_0

CHAPTER 11

1. H_0 implies die is honest
    critical χ^2 (5 df) $= 11.07$
    computed $\chi^2 = 6.2$
    $6.2 < 11.07\ \therefore$ retain H_0

3. H_0 implies 2:1 ratio of preference
    critical χ^2 (1 df) $= 3.84$
    computed $\chi^2 = \dfrac{2500}{300} + \dfrac{2500}{600} = 12.50$
    $12.50 > 3.84\ \therefore$ reject H_0, the ratio is less than 2:1.

5(a) H_0 implies conviction is not related to appearance
    critical χ^2 (2 df) $= 5.99$
    computed $\chi^2 = 7.147$
    $7.147 > 5.99\ \therefore$ reject H_0
  (b) Cramer's $\phi = .20$

7. H_0 implies that political preference is not related to year in college
    critical χ^2 (12 df) $= 12.59$
    computed $\chi^2 = 10.97$
    $10.97 < 12.59\ \therefore$ retain H_0

CHAPTER 12

1(a)

$$N = \frac{2.3^2(40 + 40)}{5^2} = \frac{5.29\,(80)}{25} = 16.93$$

(c)

$$N = \frac{2.3^2(40 + 40)}{10^2} = \frac{5.29\,(80)}{100} = 4.23$$

(e)

$$N = \frac{2.3^2(80 + 80)}{6^2} = \frac{5.29\,(160)}{36} = 23.51$$

2(2) Retained null hypothesis, so there is no need to compute η^2 and ω^2

(4) $\eta^2 = \dfrac{418.67}{468.67} = 0.89$

$\omega^2 = \dfrac{418.67 - 11.12}{468.67 + 5.56} = .86$

4(a) $S = 4$ $N = 20$ $\bar{X} = 10$ $S_{\bar{X}} = \dfrac{S}{\sqrt{N}} = \dfrac{4}{\sqrt{20}} = 0.894$

critical t_{df}, where df $= 20 - 1 = 19$; critical $t_{19} = 2.093$
$\mu = 10 \pm (2.093)(.894)$
  $= 10 \pm (1.871)$
  $= 10 + 1.871 = 11.871$
  $= 10 - 1.871 = 8.129$
$\therefore 8.129 \geqslant \mu \geqslant -11.871$

5(a)

$S = 4.037$
Critical $t_{(5)} = 2.571$ $\bar{X} = 17.5$
$\mu = 17.5 \pm (1.65)\,(2.571)$
  $= 17.5 \pm (4.242)$
  $= 17.5 + 4.242 = 21.742$
  $= 17.5 - 4.242 = 13.258$
$\therefore 13.258 \geqslant \mu \geqslant 21.742$

5(c)

$S = 11.414$
$S_{\bar{X}} = \dfrac{11.414}{\sqrt{7}} = 4.314$
$\bar{X} = 34.57$
Critical $t_{(6)} = 2.447$

$$\mu = 34.57 \pm (2.447)(4.314)$$
$$= 34.57 \pm (10.56)$$
$$= 34.57 + 10.56 = 45.126$$
$$= 34.57 - 10.56 = 24.01$$
$$\therefore 24.01 \geqslant \mu \geqslant 45.126$$

Glossary

a priori - by reason alone

absolute value - value of a number when the sign of the number is disregarded

α **error** (i.e., Type I error) - error made by rejecting a null hypothesis that is true

α **level** - probability criterion for rejecting or retaining the null hypothesis

β **error** (i.e., Type II error) - error made by retaining a null hypothesis that is false

between-subject design - design where different groups of subjects participate in the conditions representing the independent variable or variables

bivariate - something that has two variables associated with it

carryover effects - the effects of having participated in one condition carry over into another condition (carryover effects may be a problem in repeated-measure designs)

case - something that has variables associated with it (e.g., a point in a scatterplot)

categorical data - data consisting of frequency counts within predetermined categories

central tendency - the middle of a group; typical measures of central tendency are the mean, median, and mode

combination - arrangement of things where order within the arrangement is not important

computed t - value obtained from t in a t test formula

confidence interval - an interval within which a population parameter (such as the mean) will fall a certain percentage of the time

confounded experiment - an experiment that has results that can be explained by something other than the independent variable manipulation

critical t - the value that separates a t distribution into parts such that the tail or tails contain the area designated by the chosen α level; if the computed t exceeds the critical t, H_0 is rejected

decision theory - focuses on the types of errors that can be made in situations where a yes or no decision has to be made and the choice can be correct or incorrect

degrees of freedom - the number of scores within a set that are free to vary given some constraint or constraints on the set

dependent variable - variable that is influenced by an independent variable

descriptive statistics - area of statistics whose purpose is to describe a set of numbers in some summarized form

design (baseline or time series) - design wherein several measurements are taken before and after an event or manipulation has occurred; an attempt is made to establish a stable baseline before the independent variable (i.e., event) is introduced

design (pre-post) - design where a measurement is taken before and after an event has occurred; differences are attributed to the event

design (static) - design whereby subjects are placed into groups by a common characteristic

directional test - test wherein H_0 is rejected only if the outcome of a set of data is in a specified direction

empirical - through the senses (e.g., an empirical investigation depends on looking, listening, etc.)

error term - the denominator in an F test; it reflects the variability within a cell

experiment - procedure used to establish a causal link between variables

frequency distribution (cumulative) - a frequency distribution that cumulates scores going from low to high on a scale

frequency distribution (grouped) - a frequency distribution using predetermined class intervals along the x axis

frequency distribution (raw score, also called regular) - a summarized count of how many times each score within a set of scores has occurred

frequency polygon - a line graph representing frequency counts

group - condition of an independent variable that cannot be given a meaningful numbered value

histogram - displays class intervals from a frequency distribution along the x axis of a graph while the height of bars represents frequency counts

hypothesis - a statement that can be true or false

independent - not influencing or being influenced by anything else

independent groups designs - designs wherein different groups of subjects participate in each condition of the independent variable

independent variable - variable that is systematically changed, usually by an experimenter

inferential statistics - area of statistics whose purpose is to infer characteristics of a population based on samples from that population

interaction - the differential effect on a dependent variable coming from different combinations of conditions of two or more independent variables

interpolation - a procedure for determining fractional values between intervals

level - condition of an independent variable that can be given a meaningful numbered value

main effects - differences between condition means within an independent variable (e.g., between row means or between column means) in an analysis of variance

matched groups design - a design wherein each subject in one condition has a matched subject in the other conditions; matching is done with respect to the dependent variable

mean - measure of central tendency represented by $\Sigma X/N$

median - measure of central tendency that is the middle score of a ranked series of scores

mixed design - design containing both a repeated-measures factor and a between-subjects factor

modality - the number of high points in a frequency distribution

mode - measure of central tendency that is the most frequent score in a group of scores

nondirectional test - test whereby a decision about whether to reject H_0 is made independent of the direction of the outcome of the data in the test

nonparametric statistical inference - inference techniques that do not involve theoretical models or distributions; inferences are based only on random sequences of events

normal distribution - a bell-shaped theoretical distribution

null hypothesis - how characteristics of data would be expected to look in a population when the independent variable has no effect

operational definition - states the way in which an independent or dependent variable is measured

parameter - a characteristic of a population (e.g., μ)

parametric statistical inference - inference techniques that involve assumptions about population parameters and use theoretical distributions

Pearson correlation - correlational technique whereby squared distances from points in a scatterplot to a straight (i.e., regression) line are minimized

percentile - a number at or below which a certain percentage of ranked scores fall

point - a parameter such as the mean of a population

population - a group of things, individuals, or measurements that share at least one common characteristic

post-hoc tests - tests done on specific comparisons following a general test such as an analysis of variance

power - the probability of correctly rejecting H_0 in a statistical test

probability - the likelihood of an event occurring

protected t test - t tests performed following a significant F in an analysis of variance

prototype experiment - an experiment that has (a) one or more independent variables, (b) a measurable dependent variable, (c) random assignment of subjects into conditions of the independent variable

quantified - to be measured with numbered values

quasi-experiment - experiment whose design allows for explanations of the results other than the independent variable

random - by chance factors only

range - a measure of variability represented by the highest number in a set minus the lowest number in the set

regression line (also called prediction line) - best-fit straight line through a scatterplot used for predicting values of one variable based on values of the other variable

repeated-measures design - design wherein each subject participates in every condition of the independent variable

robust (as in robust test) - a statistical test that will give accurate results despite violations of the assumptions within the theoretical model underlying the test

sample - a subset of a population

sample (random) - a sample that has been selected by a method such that each member of a population has an equal chance of being selected for any designation or grouping

sample (stratified) - sample selected such that important attributes of the population are in the same proportion in the sample

scale (interval) - form of measurement that has equal intervals but no true zero point

scale (nominal) - form of measurement that only labels (e.g., numbers on uniforms)

scale (ordinal) - form of measurement that can rank but not represent equal intervals

scale (ratio) - form of measurement that has equal intervals and a true zero point

simulation - an attempt to replicate something, usually under simplified conditions

skewness - the extent to which one end of a distribution extends further from the center of the distribution than the other end

standard deviation - a measure of score spread or variability defined by $\sigma = \sqrt{\dfrac{SS}{N}}$

standard error of the difference between means - the standard deviation of a set of difference between mean (i.e., $\bar{X}_1 - \bar{X}_2$) scores

standard error of estimate - standard deviation of the distances from the points in a scatterplot to the regression line measured in units of X or Y

standard error of the mean (i.e., $\sigma_{\bar{x}}$) - standard deviation of a population of means

standard normal distribution - a distribution that results when the scores from a normal distribution are transformed into Z scores

standard score (i.e., a Z score) - a score that has been transformed by the formula $Z = (X - \mu)/\sigma$ or $Z = (X - \bar{X})/S$

statistic - a property or characteristic of a sample (e.g., the mean)

statistical model - theoretical model used to represent empirical data in making statistical inferences

T **score** - score that has been transformed by using the formula $T = 10Z + 50$

theoretical distribution - distribution defined by a mathematical formula (e.g., normal distribution or *t* distribution)

transformed score - score that has been transformed from a value on one scale to a value on a different scale through a mathematical operation

treatment effects - effects in an experiment due to an independent variable

valid measure - a measure that measures what it is supposed to measure

variable - something that can assume different values

Variance - measure of score spread defined as the square of the standard deviation

within-subject design - same as repeated-measures design

Index